Collins Revision

GCSE Higher Science

Revision Guide

FOR OCR GATEWAY B

Contents

Fit for life

Respiration

- Respiration that uses oxygen is called **aerobic respiration**.
  glucose + oxygen → carbon dioxide + water + energy

- During hard exercise not enough oxygen gets to the muscle cells. The cells also have to carry out **anaerobic respiration**. Less energy is released from anaerobic respiration.
  glucose → **lactic acid** + energy

- When lactic acid collects in muscles it causes pain and fatigue.

- The symbol equation for **aerobic respiration** is:
  $$C_6H_{12}O_6 + 6O_2 \rightarrow 6CO_2 + 6H_2O + energy$$

- During a sprint race muscles cannot get oxygen quickly enough and an **oxygen debt** builds up. Muscle cells use anaerobic respiration to release some energy.

- The energy released during anaerobic respiration is less than that from aerobic respiration because glucose is only partly broken down.

- At the end of the sprint:
  – breathing rate stays high for a few minutes to replace the oxygen
  – heart rate stays high so blood can carry lactic acid to the liver where it is broken down.

Top Tip!
Read the question carefully. It is easy to get confused between aerobic respiration and anaerobic respiration.

D–C

B–A*

Fitness

- After exercise heart rate and breathing rate take time to return to normal.
  The fitter someone is the faster they return to normal.

- However, fit people still become ill. Being fit does not stop infection by bacteria.

- Cardio-vascular efficiency, strength, stamina, agility, speed and flexibility are all ways of measuring fitness.

D–C

B–A*

Blood pressure

- Blood pressure is measured in millimetres of mercury. This is written as **mmHg**.

- Blood pressure has two measurements:
  – **systolic pressure** is the maximum pressure the heart produces
  – **diastolic pressure** is the blood pressure between heart beats.

- Diet, exercise and age can affect blood pressure.

- People with high blood pressure are often asked to fill in a questionnaire about their lifestyle.

Blood pressure questionnaire

Questions	Notes	Answers Yes	No
1 Do you take regular exercise?	Strong heart muscles will lower blood pressure		✓
2 Do you eat a healthy balanced diet?	Reducing salt intake will lower blood pressure		✓
3 Are you overweight?	Being overweight by 5 kg raises blood pressure by 5 units	✓	
4 Do you regularly drink alcohol?	A high alcohol intake will damage liver and kidneys	✓	
5 Are you under stress?	Relaxation will lower blood pressure	✓	

What changes should this person make?

- Someone with **high blood pressure** would be at increased risk from small blood vessels bursting, brain damage, strokes and kidney damage.

- **Low blood pressure** can cause problems such as poor circulation, dizziness and fainting.

D–C

B–A*

Questions

(Grades D-C)
1 Write down the word equation for anaerobic respiration.

(Grades B-A*)
2 Explain why someone breathes faster even after exercise stops.

3 Suggest what is meant by the term 'cardio-vascular efficiency'.

(Grades D-C)
4 What lifestyle factors can affect blood pressure?

What's for lunch?

Food

- Proteins from meat and fish are called **first class proteins**. They contain **amino acids** which can't be made by your body.

- In developing countries children often suffer from protein deficiency (**kwashiorkor**).

- To calculate the recommended daily amount (**RDA**) of protein, use the formula:
  RDA (in grams, g) = 0.75 × body mass (in kilograms, kg)

- A balanced diet depends on age, gender and activity.

- To calculate Body Mass Index (**BMI**) use this formula:
  $$BMI = \frac{mass\ (in\ kilograms,\ kg)}{height\ (in\ metres,\ m^2)}$$

- A BMI chart is used to see if you are the correct weight.

Look up your height and weight on the BMI bands to see where you come.

Diet

- Different people choose to have different diets for health or religious reasons.
  – Jews don't eat pork.
  – Some people have a nut allergy and mustn't eat nuts.
  – Vegetarians don't eat meat or fish and think it's wrong to kill and eat animals.
  – Vegans don't eat any foods of animal origin, including milk, cheese and eggs. They get their protein from cereals, beans, peas and nuts.

- Some people are influenced by the 'perfect' images they see in magazines. This often gives them a poor self-image and can lead to them having a poor diet.

- People with poor diets have increased health risks such as heart disease and diabetes.

Digestion

- Parts of the digestive system produce **enzymes** which break down carbohydrates, proteins and fats into smaller soluble molecules. These **diffuse** through the walls of the small intestine and into the blood plasma (carbohydrates and proteins) or lymph (fats) and pass to the cells.

Examples of different enzymes.

- The stomach makes **hydrochloric acid** to help the enzyme called pepsin work.

- Fats are difficult to digest and absorb because they are not soluble in water.

- To help with fat digestion the gall bladder produces **bile** which **emulsifies** fats. This increases their **surface area** for enzymes to act on.

Questions

(Grades D-C)
1 Calculate the RDA for someone with a body mass of 80 kg.

(Grades B-A*)
2 Suggest why magazines can give some people a poor self-image.

(Grades D-C)
3 What do digestive enzymes do?

(Grades B-A*)
4 Explain how emulsification of fats is different to digestion of fats.

Keeping healthy

Microorganisms and disease

- Mosquitoes are called **vectors** and carry microorganisms that cause malaria. Malaria is caused by a protozoan that is a **parasite** as it gets its food from its living **host**, humans. Mosquitoes pass malaria on when they bite humans.

Top Tip!

Mosquitoes carry pathogens that cause diseases; they don't carry diseases.

- Lifestyle and diet can cause different disorders.
  - A diet high in sugar often causes **diabetes**.
  - Not enough vitamin C causes **scurvy**.
  - People develop **anaemia** if they don't eat enough iron.
  - Eating a healthy diet and not smoking can reduce the risk of developing some cancers.

- Some disorders are inherited. For example, genes cause red-green colour deficiency.

- Knowledge of a vector's **life cycle** is useful in preventing the spread of disease. The mosquito larva (young stage) lives in water. Draining stagnant water kills them. Spraying **insecticide** can kill the adult mosquito. A drug called Larium can be taken by people to kill the protozoan in their blood.

- Cancer is a result of cells dividing out of control. The new cells may form tumours.
  - **Benign** tumour cells, such as in warts, are slow to divide and harmless.
  - **Malignant** tumours are cancer cells, which divide out of control and spread around the body.

Protection against microorganisms

- Symptoms of a disease are caused by the pathogen damaging cells or making **toxins**.

- Pathogens have **antigens** on their surface. When a pathogen invades the body, white blood cells make **antibodies**, which lock on to the antigens and destroy the pathogen.

- **Active immunity** happens when a pathogen invades the body a second time. The white blood cells recognise it and make antibodies quickly destroying the pathogen before the symptoms occur. Active immunity can last a lifetime.

1 this white blood cell recognises bacteria

bacteria

2 antibodies produced

- **Passive immunity** only lasts a short time and you are given antibodies in a **vaccine**.

- **Antibiotics** are drugs that attack bacteria and fungi but not viruses.

- New drugs have to be tested in different ways: on animals, specially grown human tissue or computer models. Some people object to animal testing.

3 antibodies stick bacteria together

- Each pathogen has its own set of antigens. This means that specific antibodies are needed to protect against different diseases.

4 a different type of white blood cell eats bacteria

How white blood cells work.

- Excessive use of antibiotics has lead to an increase in resistant forms of bacteria. An example is the 'superbug' MRSA.

- To test a new drug, doctors use groups of volunteers. Some take the drug and some take a harmless pill called a **placebo**. In some trials the volunteers don't know which treatment they are receiving (**blind trial**). In other trials the doctors also don't know which treatment is used (**double blind trial**); other doctors keep this information.

Questions

Grades D-C

1 Explain why mosquitoes are called vectors.

Grades B-A*

2 Describe the difference between benign and malignant tumours.

Grades D-C

3 Describe the role of antibodies.

Grades B-A*

4 Describe the difference between blind and double blind drugs trials.

Keeping in touch

Eyes

- **Binocular vision** helps us to judge distances because the range of vision from two eyes overlaps. However, it only gives a small range of vision compared to monocular vision.

- People are short- or long-sighted because their eyeballs or lenses are the wrong shape.

- About ten per cent of the population experiences problems with colour vision. Some people lack specialised cells in their retinas. This causes red-green colour deficiency, which is inherited.

- The eye focuses light by changing the size of the lens. This is called **accommodation**.

- To focus on **distant** objects, the ciliary muscles relax and tighten the suspensory ligaments. This pulls the lens, making it thin.

- To focus on **near** objects, the ciliary muscles contract the suspensory ligaments which slacken and the lens becomes fat. The fatter the lens the more it refracts the light.

- As we get older the muscles become less flexible and it becomes harder to focus.

- Glasses or contact lenses can be used to correct some eye problems.

- Cornea surgery can also correct vision. Lasers are used to change the shape of the cornea.

> **Top Tip!**
>
> Use diagrams to help your revision. Exam questions on eyes often ask you to label a diagram.

ciliary muscles control suspensory ligaments

convex lens refracts light rays

outer cornea refracts light rays

a focused image forms on the retina, which is sensitive to light

optic nerve carries nerve impulses to the brain

pupil allows light rays to enter the eye

coloured iris controls the amount of light entering the eye

suspensory ligaments alter the shape of the lens in focusing

The parts of the eye.

> **Top Tip!**
>
> A **concave** lens curves inwards like a 'cave'. A **convex** lens curves outwards.

a Short sight

← Eyeball too long →

light rays from a distant object

concave lens corrects refraction of light rays so the image lands on the retina

b Long sight

← Eyeball too short →

light rays from a distant object

convex lens corrects refraction of light rays so the image lands on the retina

Concave and convex lenses correct short and long sight.

Neurones

- **Motor neurones** carry impulses to the **muscle**.

- The nerve impulse is carried in the **axon** of the neurone.

- A **reflex arc** shows the direction in which an impulse travels: stimulus → receptor → sensory neurone → central nervous system → motor neurone → effector → response.

- Touching a hot plate is an example of a **spinal reflex arc**.

- Neurones are adapted to quickly carry and pass on nerve impulses.
  – They can be very long (nearly 2 metres).
  – They have branched endings (**dendrites**) to pick up impulses.
  – They are insulated by a fatty sheath, so that the electrical impulses don't cross over.

- Signals travel from one neurone to another across a gap called the **synapse**.

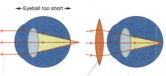

branching dendrites

muscle fibres (effector)

cell body

axon

nucleus

sheath

A motor neurone.

electrical impulse travels down the first neurone

axon

synapse

impulse triggers release of acetylcholine which diffuses across synapse

cell body of second neurone

new impulse generated in the second neurone

How a signal travels.

Questions

Grades D-C

1 Name the part of the eye that controls the amount of light entering it.

Grades B-A*

2 Explain why it is harder to focus as we get older.

Grades D-C

3 Write the following in the correct order so that they describe a reflex arc.

**effector   CNS   motor neurone   sensory neurone
receptor   stimulus   response**

Grades B-A*

4 Suggest why an impulse can only travel one way across a synapse.

Drugs and you

Drugs

D–C

- Drugs are classified by law based on how dangerous they are and the penalties for possession.

	class A	class B	class C
maximum prison sentence	7 years and fine for possession	5 years and fine for possession	2 years and fine for possession
types of drugs	heroin, methadone, cocaine, ecstasy, LSD, magic mushrooms	amphetamines, barbiturates	anabolic steroids, Valium®, cannabis

- Here are some examples of different drugs:
  - depressants: alcohol, solvents and temazepam
  - hallucinogens: cannabis and LSD
  - painkillers: aspirin and heroin
  - performance enhancers: anabolic steroids
  - stimulants: ecstasy and caffeine

B–A*

- Different people have different views on drugs. Some people in the UK believe that the personal use of drugs, such as cannabis, should be allowed. They argue that prohibition of alcohol in America in the 1930s didn't stop its use; it simply created organised crime. Other people highlight scientific studies that show how dangerous and destructive these drugs can be.

- Nicotine acts as a **stimulant**. It affects synapses (see page 7). It stimulates the acetylcholine receptors allowing more impulses to pass.

- Alcohol is a **depressant**. It affects the brain, interfering with co-ordination and balance. It binds with acetylcholine receptors, blocking nerve impulses.

Tobacco

D–C

- Cells that line the trachea and bronchioles are called **epithelial cells**. Some cells have tiny hairs called **cilia** and others make sticky mucus.

- Cigarette smoke stops the cilia from moving and dust and particulates collect and irritate the cells. Smokers cough to move this mess upwards so it can be swallowed.

The effects of smoking on the body.

B–A*

- Tars collect in air sacs and alveoli deep inside the lungs. They irritate the delicate lung tissue and are **carcinogens**.

- The tiny particulates in smoke also collect in lung tissue. They block the exchange of gases and reduce the amount of oxygen available to the rest of the body.

Top Tip!

In the exam you will need to interpret data on alcohol and cigarette smoke. Practice questions will help you with this skill.

Alcohol

Top Tip!

You need to be able to interpret information on reaction times, accident statistics and alcohol levels.

D–C

- The liver breaks down toxic chemicals such as alcohol. The liver can't deal with large quantities of alcohol. The alcohol kills the liver cells and causes **cirrhosis**.

- The alcohol content of a drink is measured in **units** of alcohol.

Questions

Grades D-C

1 Name a class A drug that is also a hallucinogen.

Grades B-A*

2 Describe the effect of a stimulant on the nervous system.

Grades D-C

3 Describe the effect of cigarette smoke on ciliated epithelial cells.

Grades B-A*

4 Describe the effect of particulates on the lungs.

Staying in balance

Homeostasis

- Various body systems keep the levels of oxygen, water, carbon dioxide and temperature constant. Keeping a constant internal environment is called **homeostasis**.

- Sweat comes from **sweat glands** in the skin and needs heat energy to evaporate. It takes heat from the body, cooling it down.
  - If your body gets too hot you could suffer from heat stroke or **dehydration**.
  - If your body gets too cold you could suffer from **hypothermia**.
  - If your body continues to get too hot or too cold you could die.

- The temperature of the body is controlled by a feedback mechanism. It is called a **negative feedback** since sweating negates (cancels out) the increasing temperature.

- The body is kept at 37 ºC because it is the **optimum temperature** for enzymes.

- The **hypothalamus** is a gland in the brain which detects the temperature of the blood. If blood temperature changes, the hypothalamus triggers protective measures such as changing the size of blood capillaries in the skin (**vasoconstriction** and **vasodilation**).

Vasoconstriction and vasodilation.

Vasoconstriction. When the body is too cold, blood capillaries in the skin constrict and so less blood flows through them, reducing heat loss

Vasodilation. When the body is too hot, blood capillaries in the skin dilate and so blood flow increases, bringing more blood to the surface where it loses heat

sweat evaporates from the skin surface, cooling it

Hormones and diabetes

- Some people have **diabetes** because their **pancreas** doesn't make enough **insulin**.

- Diabetes can be controlled by eating a diet low in sugar or by an injection of insulin. Insulin lowers the level of glucose in the blood.

- The size of the insulin dose a diabetic needs depends on their diet and exercise.

- When levels of glucose in the blood are too high, insulin converts some glucose into glycogen, which is stored in the liver.

Sex hormones

- Ovaries start producing **sex hormones** when girls reach **puberty**. The hormones cause breasts to develop and hips to widen. Hair starts to grow in the pubic area and under armpits. Periods start because the ovaries have begun to release an egg every month.

- **Testes** produce male sex hormones. Boys become more muscular, grow facial hair and their voice breaks. Their testes start to produce sperm and genitals develop.

- All these developments are called **secondary sexual characteristics**.

- **Oestrogen** and **progesterone** control the **menstrual cycle**. Oestrogen causes the repair of the uterus wall and progesterone maintains it. Together they control **ovulation**.

- Synthetic hormones can be used in contraception; they stop ovulation (egg release) by mimicking pregnancy. Female sex hormones are also used to treat infertility such as lack of eggs.

Questions

(Grades D-C)

1 Explain what is meant by the term 'hypothermia'.

(Grades B-A*)

2 Explain what is meant by the term 'vasodilation'.

(Grades D-C)

3 Describe one secondary sexual characteristic of girls and boys.

(Grades B-A*)

4 Describe the role of sex hormones in the menstrual cycle.

Gene control

Genes

- The **genetic code** is in a chemical called DNA (**deoxyribonucleic acid**). Sections of DNA form a gene.

- In DNA the two strands look like a twisted ladder. The shape is called a **double helix**.

- The rungs of the ladder are made of **four** different chemicals called **bases**. The specific arrangement of bases makes up the unique genetic code (or recipe) that makes you.

DNA.

The bases in DNA.

Cells

How the chromosome number is made.

- The number of **chromosomes** in a cell is usually an even number. This is because the paired chromosomes separate when eggs and sperm are formed.

- Different species have different numbers of chromosomes in each of their body cells. The camel has 70; the squirrel has 40; the mosquito has 6; and humans have 46.

- Eggs and sperm (**gametes**) have only one chromosome from each pair (23 in total). They combine to make a fertilised egg that has 46 chromosomes.

Top Tip!

Make sure you write a clearly explained answer to a question. Always read your answer through. If it doesn't make sense to you, it won't make sense to the examiner!

- Your body is made up of one hundred million million cells (100 000 000 000 000)!

- Each cell has a complete set of instructions contained in the DNA of its chromosomes. Even a cell in your big toe has information about your eye colour. Since this information isn't needed in your big toe cells, these genes are **switched off**.

DNA

- The four bases in DNA are shown by their initials **A**, **T**, **C** and **G**. **A** only links with **T** and **C** only links with **G**. This is important when DNA is copied because it only needs one side of the DNA ladder. The specific sequence of bases gives the genetic code.

- DNA contains codes for the production of specific **enzymes**. These in turn control cell reactions that produce specific chemicals, such as coloured eye pigments. Therefore DNA controls whether you have blue, brown or green eyes.

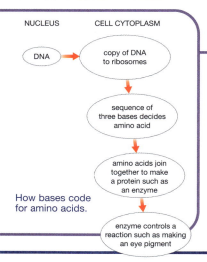

How bases code for amino acids.

Questions

Grades D-C

1 How many different bases are there in a chromosome?
2 How many chromosomes are there in the sperm cell of a camel?

Grades B-A*

3 Which base joins with base A in DNA?
4 Explain how DNA controls eye colour.

Who am I?

Male or female?

- All humans have 46 chromosomes, arranged in pairs, in their body cells.

- One pair of chromosomes is the **sex chromosome**:
  – females have an identical pair, **XX**
  – males have a non-indentical pair, **XY**.

- Each sperm carries either an X or a Y sex chromosome and has a random chance of fertilising an egg.

- There is an on-going debate over the balance between genetic (**nurture**) and environmental (**nature**) factors in determining human attributes, such as intelligence, sporting ability and health. Are great athletes made or were they born with the right genes?

a X and **b** Y sex chromosomes magnified highly. How is the shape of the X chromosome different from that of the Y chromosome?

There should be equal numbers of males and females in the human population.

Characteristics

- Humans are all different; they show **variation**. Variation is caused by genes being mixed up in gametes, fertilisation and changes in genes or chromosomes called **mutations**.

- Mutations change the DNA base sequence, stopping the cell making the protein the gene is normally coded for. They can be caused by radiation, chemicals or can occur spontaneously.

- Some mutations are harmful, such as haemophilia, while others can be advantageous.

Inheritance

- Some characteristics are **dominant** over others and more people will show the dominant characteristic than the **recessive** one. For example, lobed ears are dominant over ears with no lobes.

- **Breeding experiments** can be carried out to find dominant characteristics. The offspring (F_1 generation) of this cross all show the dominant purple colour.

- A **monohybrid cross** involves only one pair of characteristics. These are carried on a pair of chromosomes; one chromosome carrying the dominant **allele**, the other chromosome carrying the recessive allele. Alleles are different versions of the same gene. The dominant allele will always show up if the individual is **heterozygous**.

- This diagram shows how symbols are used to work out the **genetic cross** for cystic fibrosis.

- Cystic fibrosis is an inherited condition caused by a recessive gene. When both healthy parents are heterozygous for the condition (they are Cc) there is a one in four chance of their baby being **homozygous** (cc) for cystic fibrosis. Carriers of cystic fibrosis have to make important decisions, such as should they risk having children with cystic fibrosis?

Breeding green- and purple-stemmed tomato plants.

A genetic diagram for cystic fibrosis.

Questions

Grades B-A*

1 Identical twins can be used to research the debate on nature verses nurture. Suggest a reason why.

Grades D-C

2 Name one cause of a mutation.

Grades D-C

3 More people have brown eyes than blue. Which colour is recessive?

Grades B-A*

4 Draw a genetic diagram to show how cystic fibrosis is inherited from healthy parents.

B1 Summary

Diet and exercise

The **Body Mass Index** (BMI) can be used to indicate being under or over weight.

Digested food is absorbed.

Enzymes chemically digest food.

Fatigue is linked to oxygen debt.

Our energy comes from **aerobic respiration** or **anaerobic respiration**.

Anaerobic respiration releases less energy than aerobic respiration.

High blood pressure can damage the brain and kidneys.

Blood pressure has two readings:
– **diastolic** pressure
– **systolic** pressure.

A **balanced diet** contains food such as:
– **carbohydrates**
– **proteins**
– **fats**
– **nutrients** such as **minerals** and **vitamins**.

Hormones, sense organs and reflexes

Insulin controls the blood sugar level.

Endocrine glands produce hormones such as **insulin** and **sex hormones**.

Reflexes are fast reactions that do not involve the brain.

Oestrogen and **progesterone** control the **menstrual cycle**.

Neurones carry **electrical impulses**.

The body temperature of 37 °C is linked to the **optimum temperature** for enzymes.

Feedback mechanisms help to maintain a constant internal environment.

Drugs

Tobacco and alcohol are called "**social drugs**". Tobacco smoke contains many chemicals such as nicotine and tar. Alcohol affects judgement and causes liver and brain damage.

Harmful drugs are classified as class A, B and C. Any class A drug such as cocaine is dangerous and addictive.

New drugs must be tested in trials.

The mosquito is a **vector** that carries malaria.

Fungi, bacteria, viruses and protozoa can cause disease. They are **pathogens**.

Infectious diseases are easily passed on.

Our **immune system** and **immunisation** protect against infections.

Genes

Genes control the production of proteins such as enzymes.

The **coded information** is the sequence of bases in DNA.

Cystic fibrosis, red-green colour deficiency and sickle cell anaemia are inherited genetic disorders

Inherited disorders are caused by **faulty genes**.

We have **23 pairs** of **chromosomes**.

Some human characteristics are **inherited** and some are caused by the **environment**.

The **genetic code** is coded information in genes which make up chromosomes.

Cooking

Cooking food

D–C

- Some foods can be eaten **raw**, but other foods must be **cooked** to make them safer or more attractive. Food is cooked because:
  - the high temperature **kills** harmful **microbes** in food
  - the **texture** of food is improved
  - the **taste** of food is improved
  - the **flavour** of food is enhanced
  - cooked food is easier to **digest**.

Proteins

D–C

- Potatoes and flour are good sources of **carbohydrate**.

- Meat and eggs are good sources of **proteins**. Proteins are large molecules that have definite shapes. When food is cooked, the protein molecules change shape.

Proteins denature on heating.

B–A*

- The shape change is **irreversible** and the protein molecule is now **denatured**.

Baking powder

D–C

- Baking powder is a chemical called **sodium hydrogencarbonate**.

- When it's heated, it **decomposes** to give sodium carbonate, carbon dioxide and water:
  - the **reactant** is sodium hydrogencarbonate
  - the **products** are sodium carbonate, carbon dioxide and water.

- The word equation for the reaction is:

  $$\text{sodium hydrogencarbonate} \xrightarrow{\text{heat}} \text{sodium carbonate} + \text{carbon dioxide} + \text{water}$$

- It's the **carbon dioxide** in the reaction that helps **cakes** to **rise**.

B–A*

- The balanced symbol equation for this is:
  $$2NaHCO_3 \longrightarrow Na_2CO_3 + H_2O + CO_2$$

Testing for carbon dioxide

D–C

- The chemical test for **carbon dioxide** is to pass it through limewater. It turns the **limewater** from colourless to milky white.

carbon dioxide

delivery tube

limewater turns from colourless to milky white when carbon dioxide is bubbled through it

Testing for carbon dioxide.

Questions

Grades D-C
1 What do protein molecules do when they're heated?

Grades B-A*
2 When heated, proteins 'denature'. What does this mean?

Grades D-C
3 Jo says that when you heat baking powder, it decomposes and three products are made. Explain what she means.

Grades B-A*
4 Write a symbol equation for the decomposition of baking powder.

Food additives

Food additives

- Food additives are added for different reasons:
  - to **preserve** food from reacting with oxygen, bacteria or mould
  - to give a different **sensory experience**, such as to **enhance** the colour or flavour of food.

- **Antioxidants** stop food from reacting with oxygen. **Ascorbic acid** (vitamin C) is used in tinned fruit and wine as an antioxidant. Its E number is E300.

- Information about food is given on food **labels**.

D–C

Food packaging

- **Intelligent** or **active packaging** are methods used to stop food spoiling. They remove water or heat or cool the contents of packs.
  - **Active packaging** changes the condition of the food to extend its shelf life.
  - **Intelligent packaging** uses sensors to monitor the quality of the food and lets the customer know when the food is no longer fresh.

D–C

An indicator on intelligent packaging.

- **Active packaging** uses a **polymer** and a **catalyst** as a packaging film that scavenges for oxygen. It prevents the need for additives, such as antioxidants, to be added to foods. It often involves the removal of water to make bacteria or mould more difficult to grow. It's used for cheese and fruit juice.

- **Intelligent packaging** includes **indicators** on packages. An indicator on the outside of a package shows how fresh a food is. A central circle darkens as the product loses its freshness.

fresh

still fresh – consume immediately

no longer fresh

central circle darkens as food loses its freshness

B–A*

Emulsions and emulsifiers

- Detergents are long molecules made up of two parts: a head and a tail. The tail is a 'fat-loving' part and the head is a 'water-loving' part.

- Examples of **emulsions** are:
  - some **paints**
  - **milk**, which is an emulsion of oil in water
  - **mayonnaise**, which is an emulsion of oil and vinegar with egg. Egg is the **emulsifier**.

D–C

fat-loving part

water-loving part

The fat-loving part of the detergent goes into the grease droplet.

- When mayonnaise is made, the egg yolk binds the oil and vinegar together to make a smooth substance.

- The mayonnaise doesn't separate as the egg yolk has a molecule that has two parts:
  - a water-loving part that attracts vinegar to it, called the **hydrophilic head**
  - a water-hating part that attracts oil to it, called the **hydrophobic tail**.

- The hydrophobic tail is attracted into the lump of oil but the head isn't. The hydrophilic head is attracted to water and 'pulls' the oil on the tail into the water.

B–A*

water and vinegar molecules
hydrophilic head
egg yolk molecule
oil drop
hydrophobic tail
an emulsion of oil and vinegar
emulsifying molecule

An emulsifying molecule.

Questions

Grades D-C

1 What's the antioxidant added to tinned fruit and wine?

Grades B-A*

2 How does the removal of water help to preserve food?

Grades D-C

3 Emulsifiers are long molecules that have two parts. Describe the two parts.

Grades B-A*

4 Emulsifiers keep oil and water from separating. Explain how.

Smells

Esters

D–C

- To make a perfume, **alcohol** is mixed with an **acid** to make an **ester**.

  alcohol + acid ———→ ester + water

- Look at the diagram on the right.
  - Acid is added to the alcohol and is heated for some time.
  - The condenser stops the gas from escaping and helps it to cool down again.

water out

condenser

water in

ethanoic acid
ethanol and
concentrated
sulfuric acid

the perfume is
made in here

heat

Making a
perfume.

Perfume properties

D–C

- A **perfume** needs to:
  - **evaporate** easily so the perfume particles can reach the nose
  - be **non-toxic** so it doesn't poison people
  - be **insoluble** in water so it can't be washed off easily
  - not react with water so it doesn't react with perspiration
  - not irritate the skin so it can be sprayed directly onto the skin.

Solutions

D–C

- A **solution** is a solute and a solvent that don't separate.

- Esters are used as **solvents**. Other solvents can be used as cleaners, such as oil to clean grease.

- **Cosmetics** need to be thoroughly **tested** so that they don't harm humans.
  - They shouldn't cause rashes or itchiness.
  - They shouldn't cause skin damage or lead to cancer or other life-threatening conditions.

- Cosmetic testing takes many years and is highly controversial.
  - Some people object to testing on animals as the animals may be harmed, and don't have any control over what happens to them.
  - Other people say they feel safer if the cosmetics have been tested on animals.

Particles

B–A*

- If a liquid evaporates easily then the substance is **volatile**.

- **Particles** in liquid **perfume** are weakly attracted to each other. When some of these particles increase their **kinetic energy**, the force of attraction between them is overcome. The particles escape through the surface of the liquid into the surroundings as gas particles. This is **evaporation**. The gas particles move through the air by **diffusion** to reach the sensors in the nose.

- Water doesn't dissolve **nail varnish**. This is because the force of attraction between two water molecules is stronger than that between a water molecule and a molecule of nail varnish. Also, the force of attraction between two nail varnish molecules is stronger than between a nail varnish molecule and a water molecule.

Questions

Grades D-C
1 Which two substances combine to make an ester?
2 Two people may have different ideas about cosmetic testing. Explain why.

Grades B-A*
3 What does 'volatile' mean?
4 Louise removes her nail varnish with a solvent. Explain why she doesn't use water.

Making crude oil useful

Fossil fuels

- **Fossil fuels** are **finite resources** because they are no longer being made. When these fossil fuels are used up, there will be no more. They're called a **non-renewable** source.

D–C

Fractional distillation

- **Crude oil** is a mixture of many types of oil, which are all **hydrocarbons**.

- In **fractional distillation**, crude oil is heated at the bottom of a tower.
  - Oil that doesn't boil, sinks as a thick liquid to the bottom. This fraction is **bitumen** and is used to make **tar** for road surfaces. Bitumen has a very high boiling point. It 'exits' at the bottom of the tower.
  - Other fractions boil and their gases rise up the tower. Fractions with lower boiling points, such as **petrol** and **LPG**, 'exit' at the top of the tower, where it's colder.

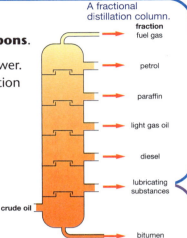

A fractional distillation column.

fraction
fuel gas

petrol

paraffin

light gas oil

diesel

lubricating substances

crude oil

bitumen

D–C

- Crude oil can be separated because the molecules in different fractions have **different length chains**. The **forces** between the molecules (**intermolecular forces**) are different and are broken during boiling. The molecules of a liquid separate from each other as molecules of gas.
  - **Heavy molecules**, such as in bitumen and heavy oil, have very long chains, so the molecules have strong forces of attraction. This means they're difficult to separate. A lot of energy is needed to pull the molecules apart. They have **high boiling points**.
  - **Lighter molecules**, such as petrol, have short chains. Each molecule has weak attractive forces and is easily separated. Less energy is needed to pull them apart. They have very **low boiling points**.

B–A*

Problems with extracting crude oil

- **Oil slicks** can harm animals, pollute beaches and destroy unique habitats. Clean-up operations are extremely expensive and the detergents and barrages used can cause **environmental problems**.

D–C

- Extracting crude oil can cause **political problems**. Oil-producing nations can set prices high and cause problems for non-oil producing nations.

B–A*

Cracking

liquid alkane on mineral fibre   aluminium oxide

Cracking.

ethene

very strong heat

water

- Cracking is a process that:
  - turns large alkane molecules into smaller alkane and alkene molecules
  - also makes useful alkene molecules with a **double bond**, which can be used to make **polymers**.

- Alkanes have a general formula of: C_nH_{2n+2}.

Octane has 8 carbon atoms and $2n + 2 = 18$ hydrogen atoms. The formula for octane is C_8H_{18}.

D–C

Questions

(Grades D-C)

1 Where do fractions with the lowest boiling points 'exit' the tower in fractional distillation?

(Grades B-A*)

2 Which products from crude oil have strong forces of attraction between the molecules? Explain why.

(Grades D-C)

3 What's the formula for an alkane with 7 carbon atoms?

(Grades B-A*)

4 How does industry match the demand for petrol with the supply from crude oil?

Making polymers

Polymerisation

- **Polymerisation** is the process in which many **monomers** react to give a **polymer**. This reaction needs **high pressure** and a **catalyst**.

- You can recognise a polymer from its **displayed formula** by looking out for these features:
  – a long chain
  – the pattern repeats every two carbon atoms
  – there are two brackets on the end with extended bonds through them
  – there's an 'n' after the brackets.

This is the displayed formula of poly(ethene).

- **Addition polymerisation** is the reaction of many monomers of the same type that have **double bonds** to form a polymer that has **single bonds**.

- The displayed formula of an addition polymer can be constructed when the displayed formula of its monomer is given.

The displayed formula for the ethene monomer is:
During a polymerisation reaction, the high pressure and catalyst cause the double bond in the ethene monomer to break and each of the two carbon atoms forms a new bond. The reaction continues until it's stopped, making a long chain. This is poly(ethene).

- If the displayed formula of an addition polymer is known, the displayed formula of its monomer can be worked out by looking at its repeated units.

This addition polymer: has a repeated unit of two carbon atoms, three hydrogen atoms and one chlorine atom. Therefore the monomer's displayed formula is:

- An **unsaturated** compound contains at least one double bond between carbon atoms. A **saturated** compound contains only single bonds between carbon atoms.

Hydrocarbons

- A hydrocarbon is a compound of carbon and hydrogen atoms only.
  – An **alkane** has a single bond C–C.   – An **alkene** has one double bond C=C.

Propane, C_3H_8, is a hydrocarbon and an alkane.

Propanol, C_3H_7OH, *isn't* a hydrocarbon because it contains an oxygen atom.

Propene is a hydrocarbon, an alkene and a **monomer**.

Polypropene is the **polymer**.

- **Alkenes** are **unsaturated**. The general formula of an alkene is: C_nH_{2n}.
  Alkanes are **saturated**. The general formula of an alkane is: C_nH_{2n+2}. It has no double bonds.

- Carbon and hydrogen atoms share an electron pair to form **covalent bonds**.

- **Bromine solution** is used to test for **unsaturation**. When an alkene is added, the orange bromine solution turns colourless because it has reacted with the alkene to form a new compound. An alkane doesn't react with bromine solution and so the bromine remains orange.

Questions

(Grades D-C)
1 What two conditions are needed for polymerisation to take place?

(Grades B-A*)
2 Look at the polymer. Draw its monomer.

(Grades D-C)
3 What's the difference between an alkane and an alkene?

(Grades B-A*)
4 What's the general formula for an alkene?

Designer polymers

Breathable polymers

Waterproof walking.

- Polymers are better than other materials for some uses.

use	polymer	other material
contact lens	wet on the eye	dry on the eye
teeth filling	attractive	looks metallic
wound dressing	waterproof	gets wet

- **Nylon** is tough, lightweight and keeps rainwater out, but it keeps body sweat in. The water vapour from the sweat **condenses** and makes the wearer wet and cold inside their raincoat.

- If nylon is **laminated** with a PTFE/polyurethane **membrane**, clothing can be made that's **waterproof** and **breathable**. Gore-Tex® has all the properties of nylon and is breathable, so it's worn by many active outdoor people. Water vapour from sweat can pass through the membrane, but rainwater can't.

- In Gore-Tex materials, the inner layer of the clothing is made from expanded PTFE (polytetrafluoroethene), which is **hydrophobic**.

- The PTFE is expanded to form a **microporous membrane**. Only small amounts of the polymer are needed to create an airy, lattice-like structure. Wind doesn't pass through the membrane.

- In expanded PTFE a membrane pore is 700 times larger than a water vapour molecule and therefore moisture from sweat passes through.

Biodegradable and non-biodegradable polymers

- Scientists are developing **addition polymers** that are **biodegradable**. These are disposed of easily by **dissolving**. Biopol is a biodegradeable plastic that can be used to make laundry bags for hospitals. It degrades when washed leaving the laundry in the machine.

- **Disposal problems** for **non-biodegradable** polymers include the following:
  - landfill sites get filled quickly and waste valuable land
  - burning waste plastics produces toxic gases
  - disposal by burning or landfill sites wastes a valuable resource
  - problems in sorting different polymers makes recycling difficult.

Stretchy polymers and rigid polymers

- The **atoms** of the monomers in each of the chains in a polymer are held together by **strong intramolecular bonds**. The **chains** in the polymer are held together by weak intermolecular forces of attraction.
  - Plastics that have **weak intermolecular forces of attraction** between the polymer molecules have **low melting points** and can be **stretched** easily as the polymer molecules can slide over one another.
  - Other plastics that form strong **intermolecular chemical bonds** or cross-linking bridges between the polymer molecules have **high melting points** and can't be stretched easily as the molecules are rigid.

monomer   intramolecular bonds are strong

intermolecular forces of attraction are weak

Intermolecular bonds are stronger than the intermolecular forces of attraction.

Questions

Grades D-C

1 What's the disadvantage of using nylon to make outdoor clothes?

Grades B-A*

2 How big are the pores in the layer in Gore-Tex® compared to the size of water vapour molecules?

Grades D-C

3 What's given off when disposing of plastics by burning?

Grades B-A*

4 Some polymers stretch easily. Explain why.

Using carbon fuels

Choosing fuels

- A fuel is chosen because of its **characteristics**.

characteristic	coal	petrol
energy value	high	high
availability	good	good
storage	bulky and dirty	volatile
toxicity	produces acid fumes	produces less acid fumes
pollution caused	acid rain, carbon dioxide and soot	carbon dioxide, nitrous oxides

- Coal produces pollution in the form of **sulphur dioxide**. This dissolves to produce **acid rain**, which damages stone buildings and statues, and kills fish and trees.

- Petrol and diesel are **liquids** so they can circulate easily in an engine, and can be stored in petrol stations along road networks.

- As fossil fuels are easy to use and the population is increasing, more are being consumed. More **carbon dioxide**, which is a **greenhouse gas**, is released and contributes to **climate change**. Many governments have pledged to try to cut carbon dioxide emissions over the next 15 years. It's a **global problem** that can't be solved by one country alone.

Combustion

- A **blue flame** means the fuel is burning in **plenty of oxygen**: **complete combustion**.

- A **yellow flame** means the fuel is burning in a **shortage of oxygen**: **incomplete combustion**.

- The word equation for a fuel burning in air is:
  fuel + oxygen ⟶ carbon dioxide + water

to filter pump to draw air through

candle

cobalt chloride paper to test for water vapour

conical flask with limewater to test for carbon dioxide

When a fuel burns in air, what's produced?

- This reaction can be shown by an experiment in the laboratory.
  – **Limewater** is used to test for carbon dioxide.
  – **White copper sulphate** powder is used to test for water, which is produced as steam.

- Complete combustion is better than incomplete combustion because: **less soot** is made, more **heat energy** is **released**, and toxic **carbon monoxide** gas isn't produced.

- A **heater** in a poorly ventilated room is burning fuel in a shortage of oxygen. It gives off poisonous carbon monoxide. Gas appliances should be checked regularly.

- Complete combustion releases useful energy. The formulae for the **products** of complete combustion are: carbon dioxide CO_2 and water H_2O.
  – **Methane** is a common hydrocarbon fuel. The formula for methane is CH_4.
  – The equation for complete combustion is: $CH_4 + O_2 \longrightarrow CO_2 + H_2O$
  The equation must be made to **balance**. There are two oxygen atoms in the reactants and three oxygen atoms in the products.
  The balanced equation is: $CH_4 + 2O_2 \longrightarrow CO_2 + 2H_2O$

Top Tip!

First count up the numbers of atoms in each molecule (shown by the subscript numbers). Don't change these numbers. Then, if necessary, add to the molecule number (large number in front of formula) to balance the numbers on either side of the arrow.

Questions

1 Explain why governments are concerned about increased carbon dioxide emissions.

2 Which liquid is used to test for carbon dioxide?

3 Complete combustion is better than incomplete combustion. Explain why.

4 Write a balanced equation for the complete combustion of pentane: C_5H_{12}.

Energy

Types of reaction

- Chemical reactions can be divided into two groups:

exothermic	endothermic
energy transferred to surroundings (energy is released)	energy taken from surroundings (absorbs energy)
temperature increases	temperature decreases
bonds made	bonds are broken
less energy needed to break bonds than make new bonds, reaction is exothermic overall	more energy needed to break bonds than make new bonds, reaction is endothermic overall

Comparing the energy from different flames

- The flame of a Bunsen burner changes colour depending on the amount of oxygen it burns in:
  – blue flames are seen when the gas burns in plenty of oxygen (**complete combustion**)
  – yellow flames are seen when the gas burns in limited oxygen (**incomplete combustion**).

- To design an experiment to compare the energy transferred in the two different flames, remember:
  – the apparatus used to compare fuels   – the amount of gas used needs to be measured
  – make the tests fair.

Comparing fuels using calculations

- The amount of energy transferred during a reaction can be calculated.
  – A spirit burner or a bottled gas burner is used to heat water in a copper calorimeter.
  – A temperature change is chosen and measured, for example 50 °C.
  – The mass of fuel burnt is measured by finding the mass before and after burning.

thermometer

copper calorimeter

known mass of water

tripod

heat

bottled gas burner

Measuring and calculating energy transferred by fuels.

- The **tests** are made **fair** by having the:
  – same mass of water   – same temperature change
  – same distance from the calorimeter to the flame.

- The tests are made **reliable** by repeating the experiment three times and excluding draughts.

- To calculate the **energy transferred**, use the following formula. The unit is **joules** (J).
  energy transferred = mass of water × 4.2 × temperature change

  > Calculate the energy transfer if 100 g of water is heated from 20 °C to 70 °C.
  > energy transferred = 100 × 4.2 × (70 – 20) = 420 × 50 = 21 000 J (21 kJ)

- To calculate the **energy output** of a fuel, use the following formula. The unit is **joules per gram** (J/g).

  $$\text{energy per gram} = \frac{\text{energy supplied}}{\text{mass of fuel burnt}}$$

  > If the water in the previous example has been heated by 3.0 g of fuel, the energy output is:
  > $$\text{energy per gram} = \frac{21\,000}{3.0} = 7000 \text{ J/g}$$

Questions

1 What's an exothermic reaction?

2 Use ideas about bond breaking and making to explain an endothermic reaction.

3 What colour flame is seen during incomplete combustion?

4 Calculate the energy per gram released by 2.0 g of fuel that raises the temperature of 100 g of water from 20 °C to 50 °C.

C1 Summary

Food and additives

All foods are chemicals. Cooking is a **chemical change**. A chemical change is **irreversible** and a **new substance is made**.

Intelligent packaging helps with storage of food.

Potatoes are **carbohydrates**. When cooked, the cell wall breaks so they are easier to digest.

Egg and meat are **proteins**. Proteins in egg and meat change shape when cooked. This process is called **denaturing**.

Food is cooked to:
– **kill microbes**
– improve the **texture**
– improve **flavour**
– make it **easier to digest**.

Additives may be added as:
– **antioxidants**
– **colours**
– **emulsifiers**
– **flavour enhancers**.

Smells and crude oil

Perfumes are **esters** that can be made from acids and alcohols.

Crude oil is separated by **fractional distillation**. The fractions with the lower boiling points exit at the top of the tower.

The larger the molecules in a fraction, the higher the boiling point. The larger the molecule, the stronger the intermolecular forces between the molecules.

Nail varnish doesn't dissolve in water. The attraction between particles in nail varnish is stronger than the attraction between water molecules and molecules in nail varnish.

There's not enough petrol made to meet the demand. There's more heavy oil distilled than needed. These larger alkane molecules can be cracked to make smaller, more useful ones, like those of petrol.

Crude oil is a fossil fuel made by dead animals being compressed over millions of years. It is **non-renewable**.

Polymers

Polymers are large, long-chain molecules made from small monomers under high pressure with a catalyst.

Nylon and Gore-Tex® can be used in coats because they are waterproof. Gore-Tex® has the advantage that it's also breathable. The material is laminated with nylon to make it stronger.

Monomers are **alkenes**, such as ethene and propene. Alkenes are hydrocarbons made of carbon and hydrogen only.

Poly(ethene) is used for plastic bags because it is waterproof and flexible. Poly(styrene) is used for packaging and insulation.

Fossil fuels and energy

If there's a good supply of oxygen, the products of **complete combustion** of a **hydrocarbon fuel** are **carbon dioxide** and **water**.

An **exothermic** reaction **transfers heat out** to the surroundings. An **endothermic** reaction **transfers heat in**.

When choosing a fuel to use for a particular purpose, several factors need to be considered:
– energy value and availability
– storage and cost
– toxicity and how much pollution they cause
– how easy they are to use.

Fuels can be compared by heating a fixed amount of water in a calorimeter and measuring the change in temperature. The energy can be calculated by the formula: energy transferred = mass of water × 4.2 × temperature change

Heating houses

Energy flow

D–C

- Energy, in the form of heat, flows from a warmer to a colder body. When energy flows away from a warm object, the temperature of that object decreases.

- Temperature is a measure of 'hotness'. It allows one object to be compared to another.

Temperature pictures

B–A*

- **Temperature** is measured on an **arbitrary** scale: you don't need to use a thermometer.

  hot ⟶ cold

- A **thermogram** uses colour to show temperature.

- **Heat** is a measurement of **internal energy** and is measured on an **absolute scale**.

Specific heat capacity

D–C

- All substances have a property called **specific heat capacity**, which is:
  - the energy needed to raise the temperature of 1 kg by 1 °C
  - measured in joule per kilogram degree Celsius (J/kg°C)
  - different for different materials.

- When an object is heated and its temperature rises, energy is **transferred**.

B–A*

- The formula for specific heat capacity is:
  energy transferred = mass x specific heat capacity x temperature change

  Calculate the energy transferred when 30 kg of water cools from 25 °C to 5 °C.
  (specific heat capacity of water = 4200 J/kg°C)
  energy transferred = mass x specific heat capacity x temperature change
  $$= 30 \times 4200 \times (25 - 5) = 30 \times 4200 \times 20$$
  $$= 2\ 520\ 000\ J = 2520\ kJ$$

Specific latent heat

D–C

- **Specific latent heat** is:
  - the energy needed to melt or boil 1 kg of the material
  - measured in joule per kilogram (J/kg)
  - different for different materials and each of the changes of state.

- When an object is heated and it changes state, energy is transferred.

B–A*

- The formula for specific latent heat is: energy transferred = mass x specific latent heat

  Calculate the energy transferred when 2.5 kg of water changes from solid to liquid at 0 °C.
  (specific latent heat of ice/water = 340 000 J/kg)
  energy transferred = mass x specific latent heat
  $$= 2.5 \times 340\ 000 = 850\ 000\ J = 850\ kJ$$

- When a substance changes state, energy is needed to break the bonds that hold the molecules together. This explains why there's no change in temperature.

Questions

Grades D-C

1 When you put an ice cube on your hand, your hand gets colder. Suggest why.

Grades B-A*

2 How does a thermogram show that there's loft insulation in the roof?

Grades D-C

3 The syrup in a steamed syrup pudding always appears to be hotter than the sponge even though they're at the same temperature. Why is this?

Grades B-A*

4 Calculate the energy transferred when 50 g of boiling water in a kettle change to steam. (Specific latent heat water/steam = 2 260 000 J/kg)

Keeping homes warm

Energy

- Different types of **insulation** cost different amounts and save different amounts of energy.

- Most energy is lost from the walls of an uninsulated house.

- To work out the most cost-effective type of insulation, the **payback time** is calculated:

$$\text{payback time} = \frac{\text{cost of insulation}}{\text{annual saving}}$$

> Draught proofing costs £80 to install and saves £20 per year on heating bills.
>
> $$\text{payback time} = \frac{80}{20}$$
> $$= 4 \text{ years}$$

roof 25%

windows 10%

walls 35%

doors 15%

floors 15%

- Different energy sources each have their advantages and disadvantages. Cost can be compared by using a consistent unit – **kWh**.

- The formula for energy efficiency is:

$$\text{efficiency} = \frac{\text{useful energy output}}{\text{total energy input}}$$

> For every 100 J of energy in coal, 27 J are transferred to a room as heat by a coal fire.
>
> $$\text{efficiency} = \frac{27}{100}$$
> $$= 0.27 \text{ or } 27\%$$

- Coal fires are very inefficient because so much heat is *lost* via the chimney.

> **Top Tip!**
>
> Remember you may have to **rearrange equations** to change the subject.

> **Top Tip!**
>
> Most of our energy sources come from burning fossil fuels. This contributes to climate change as the average home emits 7 tonnes of carbon dioxide every year.

- Energy can be transferred by:
  - **conduction** through solids, e.g. brick
  - **convection** by movement of air, e.g. in a cavity wall
  - **radiation** across space without the need for a material, e.g. foil behind a radiator.

silver foil

radiator

Would energy be saved if the aluminium foil was fixed to the front of radiators?

Questions

Grades D-C

1 Rich spends £3000 fitting double glazing. It saves him £50 each year on his heating bills. Calculate the payback time.

2 Eddie pays £600 to have foam injected into the cavity walls. He's told that the payback time is 5 years. How much can he expect to save on his heating bills each year?

3 A coal fire has an efficiency of 0.35 (35%). Mrs Tarantino spends £150 per year on coal. How much money does she waste heating the surroundings instead of the room?

Grades B-A*

4 By what process does: **a** a coal fire heat a room **b** energy transfer to the outside of a home up the chimney **c** energy transfer to the outside of a home through the windows?

How insulation works

House insulation

D–C

- **Double glazing** reduces energy loss by **conduction**. The gap between the two pieces of glass is filled with a gas or contains a **vacuum**.

- **Solids** are good conductors because the particles are close together. They can transfer energy very easily.

warm end of solid, where particles have gained energy and vibrate a lot, transferring their energy to neighbouring particles

cool end of solid, where particles are not vibrating very much

a solid is a good conductor

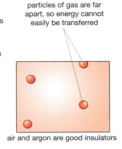

particles of gas are far apart, so energy cannot easily be transferred

air and argon are good insulators

no particles, so no energy transfer

a vacuum is the best insulator

Why is air a better insulator than a solid?

- Particles in a **gas** are far apart, so it's very difficult to transfer energy.

- There are no particles in a **vacuum** so it's impossible to transfer energy by conduction.

- **Loft insulation** reduces energy loss by **conduction** and **convection**:
  – warm air in the home rises
  – energy is transferred through the ceiling by conduction
  – air in the loft is warmed by the top of the ceiling
  – the warm air is trapped in the loft insulation
  – both sides of the ceiling are at the same temperature so no energy is transferred.

air in loft warmed by top surface of the ceiling

ceiling

warm air in room rises

Why is the ceiling a conductor?

- Without loft insulation:
  – the warm air in the loft can move by convection and heat the roof tiles
  – energy is transferred to the outside by conduction.

- **Cavity wall insulation** reduces energy loss by **conduction** and **convection**:
  – the air in the foam is a good insulator
  – the air can't move by convection because it's trapped in the foam.

- **Insulation blocks** used to build new homes have shiny foil on both sides so:
  – energy from the Sun is reflected back to keep the home cool in summer
  – energy from the home is reflected back to keep the home warm in winter.

Top Tip!

Remember, hot air will only rise into the loft if the loft-hatch is open.

Energy transfer

B–A*

- **Conduction** is due to the transfer of **kinetic energy** between particles.

- A gas expands when it's heated. This makes it less dense, so it rises.

$$\text{density} = \frac{\text{mass}}{\text{volume}}$$

- The unit of density is kg/m^3 or g/cm^3.

- **Radiation** doesn't need a material to transfer energy so energy can be transferred through a **vacuum**.

Questions

Grades D-C

1 Suggest why a vacuum-sealed double glazed unit is better than having thicker glass in a window.

2 Debbie says that snow doesn't melt as quickly on the roof of a well insulated house. Explain why.

3 Explain the advantages of having cavity wall insulation.

Grades B-A*

4 Use your ideas about energy to explain how energy is transferred from one side of a brick wall to the other.

Cooking with waves

Radiation

- **Infrared radiation** doesn't penetrate food very easily but microwaves penetrate up to 1 cm into food.

- Microwaves can penetrate glass or plastic but are reflected by shiny metal surfaces. The door of a microwave oven is made from a special reflective glass.

- **Microwave radiation** is used to communicate over long distances. The transmitter and receiver must be in **line of sight**. Aerials are normally situated on the top of high buildings.

The Telecom Tower. Why are microwave aerials on high buildings?

- **Satellites** are used for microwave communication. The signal from Earth is received, amplified and re-transmitted back to Earth. Satellites are in line of sight because there are no obstructions in space. The large aerials can handle thousands of phone calls and television channels at once.

Electromagnetic spectrum

- **Energy** is transferred by **waves**.
  - The amount of energy depends on the **frequency** or **wavelength** of the wave.
  - High frequency (short wavelength) waves transfer more energy.
  - **Gamma rays** transfer most energy and **radio waves** the least.

- **Normal ovens** cook food by **infrared** radiation. Energy is absorbed by the surface of the food so the **kinetic energy** of the surface food particles increases. The rest of the food is heated by **conduction**.

- **Microwave ovens** cook food by **microwave** radiation. The water molecules in the outer layers of food vibrate more. Energy is transferred to the rest of the food by **conduction** and **convection**.

What happens to the kinetic energy of food particles in infrared and microwave cookery?

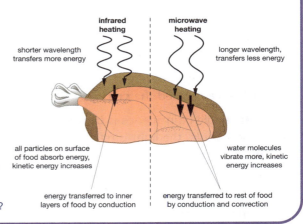

infrared heating

shorter wavelength transfers more energy

microwave heating

longer wavelength, transfers less energy

all particles on surface of food absorb energy, kinetic energy increases

water molecules vibrate more, kinetic energy increases

energy transferred to inner layers of food by conduction

energy transferred to rest of food by conduction and convection

Microwave properties

- Microwaves have wavelengths between 1 mm and 30 cm.

- **Mobile phones** use longer wavelengths than microwave ovens. This means less energy is transferred by mobile phones.

- Microwaves don't show much **diffraction** around natural objects such as hills. Therefore the signal strength for mobile phones can change a lot over a short distance.

- Mobile phones can interfere with sensitive equipment so they are banned on planes and in many hospitals.

Questions

Grades D-C

1 Suggest why microwave ovens need to have a special glass door.

2 Some areas of the country have very poor mobile phone reception. Suggest why.

Grades B-A*

3 Explain why it takes less time to cook food in a microwave oven than in a normal oven.

4 Use your ideas about frequency and wavelength to compare the heating effects from mobile phones and microwave ovens.

Infrared signals

Signals

- **Digital** signals have two values – **on** and **off**.

- **Analogue** signals can have any value and are continuously **variable**. The analogue signal changes both its **amplitude** and **wavelength**.

on (1) →
off (0) →

0 1 0 0 1 0 1 1

A digital signal.

An analogue signal.

- Before an analogue signal is transmitted, it's added to a **carrier wave**. The frequency of the carrier wave is usually higher and the combined wave is transmitted.

- **Interference** from another wave can also be added and transmitted. If the wave is amplified, the interference is amplified as well.

- Interference also occurs on digital signals, but isn't apparent because the digital signal only has two values.

- **Multiplexing** allows a large number of digital signals to be transmitted at the same time.

Total internal reflection and critical angle

- When light travels from one material to another it's **refracted**.

- If light is passing from a more dense material into a less dense material, the **angle of refraction** is larger than the **angle of incidence**.

- When the angle of refraction is 90°, the angle of incidence is called the **critical angle**.

- If the angle of incidence is **bigger** than the critical angle, the light is reflected. This is **total internal reflection**.

- Telephone conversations and computer data are transmitted long distances along **optical fibres**. Some fibres are coated to improve reflection.

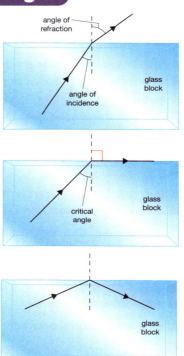

angle of refraction

glass block

angle of incidence

critical angle

glass block

glass block

- An **endoscope** allows doctors to see inside the body without the need for surgery.

- Light passes along one set of **optical fibres** to illuminate the inside of the body. The light is reflected and passes up another set of fibres to an eyepiece or camera.

Questions

Grades D-C

1 Draw a diagram of the scale of an analogue voltmeter that can read up to 5 V.

Grades B-A*

2 Explain why computer data is transmitted using digital instead of analogue signals.

Grades D-C

3 The critical angle for light passing from glass into air is 41°. Draw a diagram to show what happens if the angle of incidence is 30°.

Grades B-A*

4 Describe the advantages of using an endoscope.

Wireless signals

Radio refraction

D–C

- Radio waves are refracted in the upper atmosphere. The amount of **refraction** depends on the frequency of the wave. There's less refraction at higher frequencies.

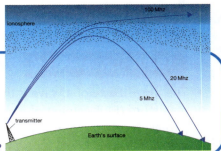

What type of wave shows most refraction?

Diffraction

B–A*

Long-wave signals diffract but short-wave and microwave signals don't.

- When waves pass through a gap they **diffract**. The amount of diffraction depends on the width of the gap compared to the wavelength. **Most diffraction** occurs when the **gap** width is **similar** to the wavelength.

- Radio waves diffract around such things as mountains. The longer the wavelength, the more the diffraction.

Reflection

B–A*

- Radio waves are reflected from the **ionosphere**. Water reflects radio waves but land mass doesn't.

- Continued reflection by the ionosphere and the oceans allows radio waves to be received from an aerial that isn't in line of sight.

- **Microwaves** pass through the ionosphere. They're received by orbiting satellites, amplified and retransmitted back to Earth.

- Communication satellites orbit above the equator and take 24 hours to orbit Earth.

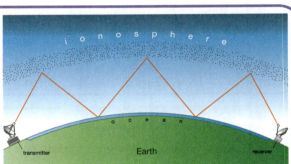

Long-wave signals diffract but short-wave and microwave signals don't.

Communication problems

D–C

- **Radio stations** broadcast signals with a particular frequency. The same frequency can be used by more than one radio station because the distance between the stations means that only one will be received. However in unusual weather conditions, the radio waves can travel further and the broadcasts **interfere**.

B–A*

- **Refraction** in the atmosphere needs to be taken into account when sending a signal to a communications satellite.

- The transmitting aerial needs to send a **focused beam** to the satellite because its aerial is very small. The transmitted beam is slightly divergent. Some energy is lost from the edge of the transmitting aerial because of **diffraction**.

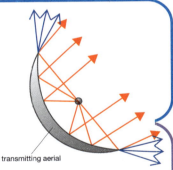

Why does a focused beam of energy need to be transmitted from a transmitting aerial?

Questions

Grades D–C

1 Radio stations that broadcast on FM use frequencies of 100 MHz or higher. Suggest why there is less interference when you listen to a radio station on FM.

Grades B–A*

2 Draw a diagram to show what happens when water waves pass through a harbour entrance that's much bigger than the wavelength of the water waves.

3 A radio transmitter is situated on the coast. A ship, below the horizon, can receive the radio broadcast clearly, but a house the same distance inland doesn't receive the signal. Suggest why.

4 Some energy is lost during a radio transmission from an aerial. Suggest why.

Light

Wave properties

- The **amplitude** of a wave is the maximum displacement of a particle **from** its rest position.

- The **trough** of a wave is the maximum displacement of a particle **below** its rest position.

- The **crest** of a wave is the maximum displacement of a particle **above** its rest position.

- The **wavelength** of a wave is the **distance between** two adjacent points of similar displacement on the wave.

- The **frequency** of a wave is the **number of complete waves** passing a point in one second.

- The formula for calculating wave speed is:
  wave speed = frequency × wavelength

The parts of a transverse wave.

When Katie throws a stone into a pond, the distance between ripples is 0.3 m and four waves reach the edge of the pond each second.
wave speed = 0.3 × 4
= 1.2 m/s

Top Tip!

Remember to always give the units in your answer:
- wavelength – **metre** (m)
- frequency – **hertz** (Hz)
- speed – **metre per second** (m/s).

Sending messages

- The **Morse code** uses a series of dots and dashes to represent letters of the alphabet. This code is used by signalling lamps as a series of short and long **flashes of light**.

- When a signal is sent by light, electricity, microwaves or radio, it's almost instantaneous.

- Each method of transmission has advantages and disadvantages:
  – Can the signal be seen by others?   – How far does the signal have to travel?
  – Can wires be cut?

- **White light** is made up of different colours of different frequencies **out of phase**.

- **Laser light** is at one frequency and is **in phase**.

How can you tell that these two waves are out of phase?

The crests and troughs occur at the same time in these two laser light waves. What term describes this?

- Laser light is used to read from the surface of a **compact disc** (CD). The surface of the CD is pitted and the pits represent a digital musical signal. Laser light is shone onto the CD surface and the difference in the reflection provides the information for the digital signal.

Questions

Grades D-C

1 A sound wave has a frequency of 200 Hz and a wavelength of 1.5 m. Calculate the speed of sound.

Grades B-A*

2 The speed of light is 300 000 km/s. The wavelength of microwaves used to transmit a particular signal is 30 cm. Calculate the frequency of the microwave transmission.

Grades D-C

3 Explain how Morse code works.

Grades B-A*

4 Draw diagrams to explain what is meant by the terms **a** in phase **b** out of phase.

Stable Earth

Earthquake waves

D–C

- The **focus** is where the earthquake happens below the surface. The **epicentre** is the point on the surface above the focus.

- **L waves** travel round the surface.

- **P waves** are **longitudinal** pressure waves:
  - P waves travel through the Earth at between 5 km/s and 8 km/s
  - P waves can pass through solids and liquids.

- **S waves** are **transverse** waves:
  - S waves travel through the Earth at between 3 km/s and 5.5 km/s
  - S waves can only pass through solids.

- A **seismograph** shows the different types of earthquake wave.

A seismograph. Which type of wave travelled fastest during the earthquake?

B–A*

- **P waves** travel through the Earth and are refracted by the **core**. The paths taken by P waves mean that scientists can work out the size of the Earth's core.

- **S waves** aren't detected on the opposite side of Earth to an earthquake because they won't travel through liquid. This tells scientists that the Earth's core contains liquid.

What do P waves help scientists find out?

What do S waves help scientists find out?

Weather effects

D–C

- **Natural events** and **human activity** affect our weather.
  - Dust from **volcanoes** reflects energy from the Sun back into the atmosphere making it cooler on Earth.
  - Dust from **factories** reflects radiation from towns back to Earth making it warmer on Earth.

- **Ultraviolet light** on the skin causes the cells to make **melanin**, a pigment that produces a tan. People with dark skin don't tan easily because ultraviolet radiation is filtered out.

- Use a sunscreen with a high **SPF** (sun protection factor) to reduce risks.
  The formula for the safe length of time to spend in the sun is:
  safe length of time to spend in the Sun = published normal burn time × SPF

B–A*

- **Ozone** is found in the **stratosphere**. Ozone helps to filter out ultraviolet radiation.

- **CFC gases** from aerosols and fridges destroy ozone and reduce the thickness of the ozone layer.

Questions

Grades D-C

1 What's the difference between the focus and the epicentre of an earthquake?

Grades B-A*

2 How do seismic waves provide information about the internal structure of Earth?

Grades D-C

3 Khan sees that on a hot day the normal burn time is 5 minutes. How long can he stay in the sun, without burning, if he uses sunscreen with SPF 30?

Grades B-A*

4 What will be the long term effects if we don't reduce the amount of CFCs we discharge into the atmosphere?

P1 Summary

Heat and temperature

Energy is **transferred** from a hotter to a colder object.

Energy is **transferred** when a substance **changes state**. The amount of energy transferred depends on:
– the mass
– the specific latent heat.

$$\text{energy transferred} = \text{mass} \times \text{specific latent heat}$$

Temperature is a measure of **hotness** in °C.
Heat is a measure of **energy transfer** in J. Energy is **transferred** when a substance **changes temperature**. The amount of energy transferred depends on:
– the mass
– temperature change
– specific heat capacity.

$$\text{energy transferred} = \text{mass} \times \text{specific heat capacity} \times \text{temperature change}$$

Energy transfer

$$\text{efficiency} = \frac{\text{useful energy output}}{\text{total energy input}}$$

Air is a good **insulator** and reduces energy transfer by **conduction**. Conduction in a solid is by the transfer of **kinetic energy**.

Trapped air reduces energy transfer by **convection**.

Shiny surfaces **reflect** infrared radiation to reduce energy transfer.

Energy saving in the home can be achieved by:
– double glazing
– cavity wall insulation
– draught strip
– reflecting foil
– loft insulation
– curtains
– careful design.

Waves carrying energy

Warm and hot objects emit **infrared radiation**. Infrared radiation is used for cooking.

Microwaves can be used for **cooking** and for **communication** when transmitter and receiver are in line of sight.

Digital and **analogue signals** are used for communication.
– Digital signals allow many signals to be transmitted at the same time and are clearer.

Radio waves are used for **communication**. Longer wavelengths diffract around obstacles.

Waves in the **electromagnetic spectrum** are:
– **radio waves**
– **microwaves**
– **infrared**
– **visible light**
– **ultraviolet**.
All electromagnetic waves can be reflected and refracted. The energy of the wave increases as the wavelength decreases.

The stable Earth

Earthquake waves travel through the Earth. Different waves help us find out about the inside of the Earth.

Exposure to **ultraviolet radiation** causes **sun burn** and **skin cancer**. Sunscreen and sunblock reduces damage caused by ultraviolet waves. CFCs are causing the ozone layer to become thinner.

Global warming is a result of both **human activity**, such as burning fossil fuels, and **natural phenomena**, such as volcanoes.

Ecology in our school grounds

Ecosystems

D–C

- We know more about the surface of the Moon than we know about the deepest ecosystems of our oceans.

- Animals from the deep can't live near the surface and are rarely seen. For example, the giant squid can grow up to 20 m long but suffocates in warm surface water. To see it in its natural habitat requires very expensive submarines. A human would be crushed due to the increased pressure at such depths.

The giant squid.

- Many new species may exist at depths that humans can't reach.

B–A*

- Oceans are natural ecosystems; they exist without any help from humans.

- There are many artificial ecosystems controlled by humans. A field of wheat is mainly artificial. Farmers try to control what grows and lives there.
  - They use **herbicides** to remove weeds.
  - They use **pesticides** to control pests.
  - They can increase the crop yield by adding **fertilisers**.

- **Biodiversity** describes the range of living things in an ecosystem. The farmer decreases the biodiversity of his farm when he uses herbicides and pesticides.

Counting animals

D–C

- To estimate a **population**, scientists can use a method called 'mark and recapture'.
  - The animals are trapped and marked in some harmless way.
  - They are then released and the traps are set again a few days later.

- To estimate the population the following formula is used.

$$\frac{\text{number of animals caught first time} \quad \times \quad \text{number of animals caught second time}}{\text{number of marked animals caught second time}} = \text{population}$$

> **Top Tip!**
>
> A **population** is a group of animals or plants of the same species. A **community** is lots of different species living in the same ecosystem.

- Population sizes will always be changing because:
  - animals are being born and others are dying
  - there is movement of animals in and out of the ecosystem.

B–A*

- To estimate a plant population, scientists can use **quadrats**.
  - The quadrat is put on the ground and the percentage cover of each plant is recorded.
  - The quadrat is placed in a random way to ensure a fair representation of an area.

- To increase the accuracy of an estimate:
  - the process is repeated several times
  - the sample size is as large as possible.

- Scientists have to remember that their samples may be unrepresentative of the population as a whole.

Questions

(Grades D-C)

1 What happens to a giant squid in warm surface water?

(Grades B-A*)

2 Name the type of chemical used to remove
   a weeds **b** pests.

(Grades D-C)

3 What is meant by the term 'population'?

(Grades B-A*)

4 Describe one method used to place quadrats in a random way.

Grouping organisms

Plants and animals

- Vertebrate animal groups have different characteristics:
  - **Fish** have wet scales and gills to get oxygen from the water.
  - **Amphibians** have a moist permeable skin.
  - **Reptiles** have dry scaly skin.
  - **Birds** all have feathers and beaks, but not all of them can fly.
  - **Mammals** have fur and produce milk.

Top Tip!

You may be asked about a vertebrate you have never heard of. Don't worry, it will fit into one of the five groups

- This table shows the main differences between plants and animals.

	food	shape	movement
plants	make their own food using chloroplasts to trap energy from the Sun	spread out to collect plenty of water and nutrients from soil	stay in one place, they cannot get up and move somewhere else
animals	cannot make their own food, so they need to eat	more compact than plants to help them move around	able to move around to find food

D–C

Odd ones out

- Some organisms don't fit into the animal kingdom or the plant kingdom.
  - Mushrooms cannot move like an animal or make food like a plant. They belong in the kingdom called **fungi**.
  - *Euglena* can make their own food or they can feed. They belong in the kingdom called **protoctista**.
  - *Archaeopteryx* has bird and reptile characteristics. It is thought to represent the evolutionary link between reptiles and birds.

- Dolphins and whales are special animals. They are mammals, yet they have **evolved** to look like fish so they can live in water.

- **Hybrids** are the result of breeding two animals from different species, such as the donkey and the horse to produce a mule. Hybrids are sterile and can't breed. It's not a species and therefore it's difficult to classify.

B–A*

What is a species?

- A tiger and a lion are both cats. However, they are two different **species**.

- Members of the same species can breed. Lions breed with other lions to make young lions. These young lions will be **fertile**.

- All cats belong to the same **family**. The family is called Felidae. Each species of cat is given its own scientific name. Lions have the name *Panthera leo*. Tigers are called *Panthera tigris*. They both have the same first name because they are closely related.

- The lion and the tiger share a recent ancestor but have evolved to live in different **habitats**. Lions live in open grassland while tigers prefer forest.

- Cheetahs and leopards aren't closely related. However, they are very similar because they are **adapted** to the same habitat.

D–C

Questions

Grades D–C

1 Explain why plants need chloroplasts.

3 Explain why dolphins need to look like fish.

Grades B–A*

2 Explain why mushrooms are not classified as plants.

Grades D–C

4 What is the family name for cats?

The food factory

Photosynthesis

D–C

- Photosynthesis can be described using this word equation:

$$\text{carbon dioxide} + \text{water} \xrightarrow[\text{(chlorophyll)}]{\text{(light energy)}} \text{glucose} + \text{oxygen}$$

- The products of photosynthesis are **glucose** and **oxygen**.

- Glucose is transported as **soluble** sugars to parts of the plant where it is needed.

- The glucose can be used by the plant in these ways:

How the products of photosynthesis are used.

B–A*

- The overall balanced chemical equation for photosynthesis is:

$$6CO_2 + 6H_2O \xrightarrow[\text{(chlorophyll)}]{\text{(light energy)}} C_6H_{12}O_6 + 6O_2$$

- For storage, the glucose joins together to make larger molecules of starch. Starch is:
  – insoluble and therefore not easily lost from the cell in solution
  – not very reactive, making it a good storage molecule.

Increasing and limiting photosynthesis

D–C

- The following actions increase the rate of photosynthesis:
  – keeping plants warm
  – giving plants extra carbon dioxide
  – increasing the amount of light.

B–A*

- This graph shows the effect of light intensity on the rate of photosynthesis:
  – the rate of photosynthesis increases as the light intensity increases until point B
  – at point B, the rate of photosynthesis stays the same, something else is limiting it
  – the limiting factor could be carbon dioxide level or temperature
  – carbon dioxide, light intensity and temperature are all **limiting factors** of photosynthesis.

Why does the curve in the graph flatten out?

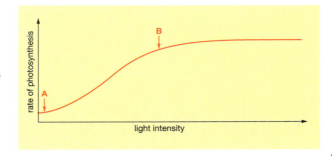

Respiration and gas exchange

D–C

- **Respiration** uses oxygen to release energy from glucose. At the same time it releases carbon dioxide and water.

B–A*

- As long as a plant is photosynthesising it needs to take in carbon dioxide. At the same time it will release oxygen.

- At night the plant still needs oxygen for respiration. It takes in oxygen from the air and releases carbon dioxide.

Top Tip!

Plants carry out respiration 24 hours a day, not just at night. If they stopped respiring they would have no energy and would die.

Questions

(Grade D-C)
1 Name three uses of glucose in a plant.

(Grade B-A*)
2 Write out the balanced equation for photosynthesis.

(Grade D-C)
3 Name three actions that increase the rate of photosynthesis.

(Grade B-A*)
4 Explain why plants give out carbon dioxide at night.

Compete or die

Plant and animal competition

- Bluebells flower in spring to catch as much light as possible before the leaves are fully out on the trees, causing shade. In summer the bluebells find it difficult to grow as the larger trees take most of the light, water and minerals. The bluebells and trees are in **competition** with each other.

- Animals often compete to attract a mate so that they can breed. Animals need to breed so that the species survives. Male elephant seals fight each other in order to keep their mates.

> **Top Tip!**
> If you're asked about animals or plants you have never heard of, don't be put off.
> All animals and plants compete for similar things.

- All organisms have a role to play in an ecosystem. For example, the role of a squirrel is to live in woods and eat acorns. This role is called the squirrel's **ecological niche**. In Britain, there are two types of squirrel: the red squirrel and the American grey squirrel. At one time they both occupied the same niche, but in different countries. Now they compete for the same niche in Britain. The grey squirrel is 'outcompeting' the red squirrel.

Animal relationships

- There are many ways in which animals of different species interact.
  - Lions eat antelopes. If there are lots of lions, antelope numbers will go down as more get eaten. When there are fewer antelopes, lion numbers go down as there is not enough food.
  - The tapeworm is a **parasite**. It lives in the digestive system of other animals including humans. The tapeworm takes food away from its **host** so that it can grow.
  - The sharksucker is a fish that attaches itself to sharks. It cleans the shark's skin by eating its parasites. In return, the shark protects the sharksucker from predators. Relationships like this where both animals benefit is called **mutualism**.

- Some species are totally dependent on others. The pea plant is a **legume**. It has structures on its roots called **root nodules**. Bacteria live inside the nodules and convert nitrogen into nitrates. They are called **nitrogen-fixing bacteria**. The bacteria give the pea plant extra nitrates to help it grow and the pea plant gives the bacteria sugar which they turn into energy.

Populations and cyclic fluctuation

- Predator–prey relationships play an important part in controlling populations.

- The 'up and down' pattern of population change is called **cyclic fluctuation**.
  This graph shows the cyclic fluctuation of the snowy owl population and its prey, the lemming.

Why are the peaks in the snowy owl population slightly behind the peaks in the lemming population?

Questions

Grades D-C

1 Explain why animals need to compete for a mate and breed.

Grades B-A*

2 The red squirrel population of Britain has gone down. Suggest a reason.

Grades D-C

3 What is meant by the term 'mutualism'?

Grades B-A*

4 Explain why the snowy owl population rises when lemming population rises.

Adapt to fit

Animals and adaptation

- Camels live in the **desert**. They are well adapted to survive the heat and lack of water.
  - All their body fat is in the hump, so heat can be lost from the rest of the body.
  - Stored fat in the hump can be used when there is no other food available.
  - Their body temperature can rise above normal, so they don't need to sweat.
  - Bushy eyelashes and hair-lined nostrils stop sand getting in.
  - Large feet spread out their weight to stop them sinking into the sand.

- Polar bears live in the **Arctic**. They are well adapted to the cold.
  - They have thick white fur for camouflage and insulation.
  - A layer of fat over its body called **blubber** keeps it insulated.
  - A large body compared to its surface area stops it losing too much heat.
  - Small ears reduce the surface area from which heat can be lost.
  - Sharp claws and teeth help it to seize and eat prey.
  - Strong legs for running and swimming.
  - Large feet spread its weight on the snow.
  - Fur-covered soles on its paws help it to grip and insulate from the cold.

- Brown bears and polar bears live in different habitats. Brown bears would find it difficult to exist in the polar bear's habitat because they aren't adapted for the cold.

Top Tip!

Make sure you know the different meanings of 'describe' and 'explain'.
- The camel has large feet – is to 'describe' an adaptation.
- The camel has large feet to spread the load on sand – is to 'explain' an adaptation.

Plants and adaptation

- **Cacti** are plants that live in the desert. They are well adapted to hot dry conditions.
  - They have long roots to reach as much water as possible.
  - A thick waterproof **cuticle** reduces water loss.
  - A fleshy stem stores water.
  - Leaves have become spines to reduce water loss and to stop animals getting at the water in the stem.
  - Photosynthesis takes place in the green stem.
  - Round shape reduces the plant's **surface area**, cutting down water loss.

- The cactus is adapted to the desert and wouldn't survive in Britain as there's too much water which would cause it to rot.

- In order to reproduce, plants need to transfer pollen from one plant to another. The process they use is called **pollination**. Plants have adapted to carry out pollination in two ways.
  - Wind pollinated plants have feathery stigmas and small, light pollen.
  - Animal pollinated plants have colourful petals, nectar and 'sticky' pollen.

Questions

Grades D-C

1 Explain why camels have bushy eyelashes.
2 Describe how polar bear feet are adapted to life in the Arctic.

Grades B-A*

3 Explain why a cactus needs a thick cuticle.
4 Suggest a reason why some plants make nectar.

Survival of the fittest

Fossils

- Fossils form in different ways.
  - Hard parts such as shells and bones can be replaced by minerals, which turn to stone.
  - Some organisms sink into mud and then **casts** or **impressions** form when they decay.
  - Organisms can be preserved in **amber**, peat bogs, tar pits or ice.

- The **fossil record** shows how organisms have changed. Not all living things have a complete fossil record, because:
  - some body parts decay quickly before they can be fossilised
  - fossilisation is rare and most living things will completely decay
  - there may still be fossils we have not found.

How the shape of the horse's foot has evolved over time.

Interpretation of the fossil record

- Some scientists use the fossil record to show how animals and plants have evolved. Other scientists have used the gaps in the fossil record to argue against evolution.

- Many complex organisms in the fossil record appear and then disappear which **Creationists** interpret to mean that organisms were created and did not evolve.

Natural selection

- Environments can change. To survive these changes a species needs to adapt and evolve otherwise it will become **extinct**. In any species, it's only the best adapted that survive.

- This **survival of the fittest** is called **natural selection**. **Genes** pass on the successful characteristics of a species to the next generation. These are examples occurring today.
  - Peppered moths are dark or pale in colour. Dark moths are better camouflaged in polluted areas, so more of them survive.
  - Rats have evolved to become resistant to the poison warfarin.
  - Bacteria are becoming resistant to antibiotics.

Top Tip!

You need to be able to use natural selection to explain how bacteria have become resistant. Remember, it's all about survival of the fittest and passing on genes.

Evolution

- Charles **Darwin** developed the theory of natural selection.
  - Within any species there is **variation**.
  - There will be competition for limited resources such as food.
  - Only those best adapted will survive, called survival of the fittest.
  - Successful adaptations are passed to the next generation in genes.
  - Over time, the changes may result in a new species.
  - The less well adapted species may become extinct.

- Jean Baptiste de **Lamarck** had a different theory called the law of **acquired characteristics**. His theory was discredited because acquired characteristics cannot be passed on by genes.

Questions

1 Write down one reason why the fossil record is incomplete.

2 Explain how Creationists interpret gaps in the fossil record.

3 How do rats pass on resistance to warfarin to the next generation?

4 Write down the main points in Darwin's theory of evolution.

Population out of control?

Pollution

D–C

- Increasing levels of carbon dioxide cause **climate change** and sulphur dioxide causes **acid rain**.

- An increase in the use of chemicals called CFCs has led to a depletion of the ozone layer.

Top Tip!
Try not to mix up the three main effects of pollution. It's CFCs that deplete the ozone layer, *not* carbon dioxide.

The main effects of pollution.

Population and pollution

B–A*

- The graph shows the past, present and predicted future world human population.

- The human population is growing at an ever-increasing rate. This is called **exponential growth**.

- The increasing population is quickly using up the Earth's resources and increasing pollution.

- The world population figures show the greatest rise in population is occurring in under-developed countries such as Africa and India.

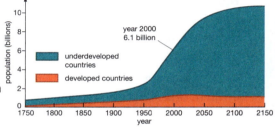

A graph showing exponential growth.

- However, if the countries that use the most fossil fuels are considered, developed countries such as the United States of America and Europe seem to be causing the problem.

- America is the heaviest user of oil, using about 50 L per person each day.

Indicator species

D–C

- The presence or absence of an **indicator species** is used to estimate levels of pollution.
  - The stonefly larva is an insect that can only live in clean water.
  - The bloodworm, water louse, sludge worm and rat-tailed maggot are animals that can live in polluted water.
  - Lichen grows on trees and rocks but only when the air is clean. It is unusual to find lichen growing in cities. This is because it is killed by the pollution from cars.

Top Tip!
You only need to remember that indicator species are used. You don't have to remember the level of pollution that each species tolerates.

B–A*

- Animals have different sensitivities to environmental conditions. In rivers and ponds, different animals can tolerate different pollution levels.
  - The sludge worm lives in polluted water because it can cope with very low oxygen levels.
  - The alderfly cannot live in polluted water. It cannot tolerate low oxygen levels.

- Water that contains lots of different species (higher biodiversity) is usually a healthy environment.

Questions

Grades D-C

1 Which gas causes acid rain?

Grades B-A*

2 Explain why the USA causes more pollution than Ethiopia.

Grades D-C

3 Name two indicator species that can live in polluted water?

Grades B-A*

4 Explain why the alderfly cannot live in polluted water.

Sustainability

Extinction

- Animals become **endangered** or **extinct** because:
  - their climate changes and they cannot adapt fast enough
  - their habitat is destroyed or becomes polluted
  - they are hunted by humans or outcompeted by better adapted animals.

- Animals can be saved from extinction by:
  - protecting habitats or setting up **artificial ecosystems** and **captive breeding** in zoos
  - making hunting illegal and educating people about the reasons to save them.

D–C

Conservation

- Conservation programmes such as saving the rainforest are important because:
  - they protect plants and animals that could be used for food or medicine
  - protecting species low down in the food chain helps those higher up to survive
  - resulting tourism can benefit local communities, helping them to maintain their culture.

B–A*

Whales

Whale parts have many uses.

- Live whales are important as people can make money from the tourist trade.

- Some whales are kept in captivity for research, breeding programmes or entertainment. Many people object when whales lose their freedom.

skin: used in belts, shoes, handbags and luggage

sinews: used in tennis rackets

spermacetti: used in high-grade machine oil

oil: sperm whale oil taken from bone and skin used in high-grade alcohol, shoe cream, lipstick, ointment, crayons, candles, fertiliser, soap and animal feeds

whalemeat: used in pet food and human food

liver: used in oil

bone: used in fertiliser and animal feeds

teeth: used in buttons, piano keys and jewellery

ambergris: from intestine, used in perfumes

- Some countries want to lift the ban on whaling. Scientists need to kill some whales to find out how they survive at extreme depths. Migration patterns and whale communication can only be investigated if the animal is alive.

D–C

B–A*

Sustainable development

- **Sustainable development** is a way of taking things from the environment but leaving enough behind to ensure a supply for the future and to prevent permanent damage.
  - Fishermen have then been set **quotas** so that there are enough fish left to breed.
  - People are educated about the importance of maintaining species' numbers for future generations.
  - Woods are re-planted to keep up the supply of trees.

- As the world population increases, it is even more important to carry out sustainable development on species such as whales.

- The demand for food and other resources could lead to an increase in whaling. The whaling nations will need to work together to prevent extinction.

- When whaling quotas are set, other factors will need to be taken into account such as pollution levels and over-fishing of the whales' food.

D–C

B–A*

Questions

(Grades D-C)

1 Explain why 'climate change' could lead to extinction of some species.

(Grades B-A*)

2 Suggest a reason how tourism could help maintain local culture.

(Grades D-C)

3 Suggest a reason why people object to keeping whales in captivity?

4 Suggest one way Britain could help sustain the rainforest in Brazil.

B2 Summary

Ecosystems

Different counting methods are used to estimate populations.
These estimates are often inaccurate if the sample size is not large enough.

All the living things and their surroundings make up an ecosystem.

Keys can be used to identify the animals and plants in a habitat.

Artificial ecosystems are often controlled using pesticides, herbicides and fertilisers.

Many ecosystems are still unexplored and could contain new species.

Classification

The **animal kingdom** is split into two groups:
– **vertebrates**
– **invertebrates**.

Plants can make their own food by a process called photosynthesis:

carbon dioxide + water $\xrightarrow[\text{(chlorophyll)}]{\text{(light energy)}}$ glucose + oxygen

$$6CO_2 + 6H_2O \xrightarrow[\text{(chlorophyll)}]{\text{(light energy)}} C_6H_{12}O_6 + 6O_2$$

There are five vertebrate groups:
– **fish**
– **amphibians**
– **reptiles**
– **birds**
– **mammals**.

Competition and survival

Whales could still be hunted to extinction. Their population needs to be sustained.

Animals and plants adapt to their **habitats**. Those better adapted are more able to **compete** for resources. Species that cannot adapt may become **extinct**. **Fossils** provide evidence of extinct species.

The **survival** of a species depends on how well it can adapt to changes in the environment. The number of predators can affect population numbers of prey and vice versa.
Survival may also depend on the presence of another organism:
– **mutualism**
– **parasitism**.

Indicator species are used to monitor pollution levels.

The population of predators and prey regulate themselves, this is called **cyclic fluctuation**.

The increase in human population is leading to an increase in pollution and loss of habitat. As habitats become smaller species are unable to compete and become extinct.
Species can be protected from extinction if resources are carefully managed – **sustainable development**.

Paints and pigments

Paints and pigments

- A paint is a **colloid** where small solid particles are dispersed through the whole liquid, but aren't dissolved.

- When an **oil paint** is painted onto a surface, the solvent **evaporates** leaving the binding medium to dry and form a skin, which sticks the pigment to the surface.

- An **emulsion paint** is a water-based paint. It's made of tiny droplets of one liquid in water, which is called an **emulsion**. When emulsion paint has been painted on to a surface as a thin layer, the water evaporates leaving the binding medium and pigment behind. As it dries it joins together to make a continuous film.

- Oil paint and emulsion paints are colloids because they are a mixture of solid particles in a liquid. The particles don't separate because they're scattered throughout the mixture and are small enough not to settle at the bottom.

- The oil in oil paint is very sticky and takes a long time to harden. Once the solvent has evaporated the oil slowly reacts with oxygen in the air to form a tough, flexible film over the wood. The oil binding medium is **oxidised** by the air.

D–C

B–A*

Thermochromic pigments

- **Thermochromic pigments** are used in paints that are chosen for their colour and also for the temperature at which their colour changes. For example, a thermochromic pigment that changes colour at 45 °C can be used to paint cups or kettles to act as a warning.

- Most thermochromic pigments change from a colour to colourless.

- Thermochromic paints come in a limited range of colours. To get a larger range of colours they are mixed with different colours of normal acrylic paints in the same way that you mix any coloured paints.

- When a green mixture gets hot, the blue thermochromic paint becomes colourless, so all that is seen is the yellow of the acrylic paint.

D–C

B–A*

Phosphorescent pigments

- **Phosphorescent pigments** absorb energy and store it. They can then release it slowly as light. They are sometimes used in luminous clock dials.

- Phosphorescent paints are much safer than radioactive paints that were developed to glow in the dark.

B–A*

Questions

(Grades D-C)
1 What is a colloid?

(Grades B-A*)
2 Describe how oil paint dries and hardens.

(Grades B-A*)
3 A company is concerned with the type of fumes given off during the drying process when painting. Which type of paint would you recommend, emulsion or oil-based. Explain your answer.

Construction materials

The raw materials

D–C

- Limestone is easier to cut into blocks than marble or granite. Marble is much harder than limestone. Granite is harder still and is very difficult to shape.

- Brick, concrete, steel, aluminium and glass come from materials in the ground, but they need to be manufactured from raw materials. This is shown in the table.

raw material	clay	limestone and clay	sand	iron ore	aluminium ore
building material	brick	cement	glass	iron	aluminium

B–A*

- **Igneous** and **metamorphic** rocks are normally harder than **sedimentary** rocks.
  - Granite is an **igneous rock** and is very hard. Igneous rock is formed out of liquid rock that cools slowly and forms interlocking crystals as it **solidifies**. It is this interlocking structure that gives the rock its hardness.
  - Marble is a **metamorphic rock**. It is not as hard as granite but is harder than limestone. Metamorphic rocks form when a rock has been under **heat and pressure** in the Earth's crust, making it harder than the original rock.
  - Limestone is a **sedimentary rock** and is the softest. Sedimentary rock is made from the shells of dead sea-creatures that stuck together.

Cement and concrete

D–C

- **Thermal decomposition** is the chemical breakdown of a compound into at least two other compounds under the effect of heat.

- Calcium carbonate (limestone) thermally decomposes at a very high temperature. This is shown in the word equation:

  calcium carbonate ⟶ calcium oxide + carbon dioxide

- **Cement** is made when limestone and clay are heated together.

- Reinforced concrete has steel rods or steel meshes running through it and is stronger than concrete. It's a **composite** material.

B–A*

- The symbol equation for the thermal decomposition of calcium carbonate is:

  $$CaCO_3 \longrightarrow CaO + CO_2$$

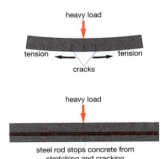

steel rod stops concrete from stretching and cracking

- Reinforced concrete is a better construction material than non-reinforced concrete.
  - If a heavy load is put on a concrete beam it will bend very slightly. When a beam bends its underside starts to stretch, which puts it under tension and cracks start to form.
  - Steel is strong under tension. Steel rods in reinforced concrete stop it stretching.

(Grades D-C)
1 What is cement made from?

(Grades B-A*)
2 How is limestone changed into marble?

(Grades D-C)
3 What are the products of heating limestone?

(Grades B-A*)
4 Write the symbol equation for the thermal decomposition of calcium carbonate.

Does the Earth move?

The structure of the Earth

- The outer layer of the Earth is called the **lithosphere**. It's relatively cold and rigid and is made of the crust and the part of the mantle that lies just underneath.

- The **tectonic plates** that make up the Earth's crust are **less dense** than the mantle and they 'float' on it. There are two kinds of plate:
  - **continental plates** that carry the continents
  - **oceanic plates** that lie underneath the oceans.

The parts of the lithosphere.

- The crust is far too thick to drill through, so most of our knowledge comes from measuring seismic waves produced by earthquakes. This technology improved in the 1960s when scientists were developing ways of detecting nuclear explosions.

D–C

- **Tectonic plates** move very slowly and in different ways: **apart**, **collide**, or **scrape sideways** past each other.

- The **mantle** is hard and rigid near to the Earth's surface and hotter and non-rigid near to the Earth's core. The mantle is always solid, but at greater depths it's more like Plasticine, which can 'flow'.

- In plate tectonics, energy from the hot core is transferred to the surface by slow **convection currents** in the mantle. Oceanic plates are denser than continental plates.
  - When two plates collide, the more dense oceanic plate sinks below the less dense continental plate. This is called **subduction**. The oceanic plate partially re-melts and is reabsorbed into the mantle.

- There have been developments in the theory of plate tectonics.
  - People noticed that the coastline of Africa matches that of South America, suggesting that the continents were formed from one 'supercontinent' which was splitting apart at the time of the dinosaurs (**continental drift theory**).

B–A*

Top Tip!

Subduction zones are where plates are being destroyed.

Magma, rocks and volcanoes

- Magma rises through the Earth's crust because it's less dense than the crust. It cools and solidifies into **igneous** rock either after it comes out of a volcano as lava, or before it even gets to the surface.

D–C

- By looking at crystals of **igneous rock**, geologists can tell how quickly the rock cooled.
  - Igneous rock that **cools rapidly** (close to the surface) has small crystals.
  - Igneous rock that **cools slowly** (further from the surface and better insulated) has large crystals.

- Different types of magma can cause different types of eruption at the surface of the Earth.
  - **Iron-rich magma** (called basaltic magma) is runny and fairly 'safe'. The lava spills over the edges of a volcano and people can escape.
  - **Silica-rich magma** is less runny and produces volcanoes that may erupt explosively. It shoots out as clouds of searingly hot **ash** and **pumice**. The falling ash often includes large lumps of rock called **volcanic bombs**.

B–A*

- Geologists investigate past eruptions by looking at the ash layers. In each eruption, coarse ash falls first, then fine ash, producing **graded bedding**. Future eruptions can sometimes be predicted with the help of seismometer, but it isn't precise and disasters still occur.

Questions

(Grades D-C)
1 Why do tectonic plates 'float' on the molten rock?

(Grades B-A*)
2 Describe the process of subduction.

(Grades D-C)
3 Explain why the size of crystals changes when magma cools.

(Grades B-A*)
4 Explain why silica-rich magma is more dangerous than iron-rich magma.

Metals and alloys

Electrolysis

D–C

- Impure copper can be purified in the laboratory using an **electrolysis cell**.
  - The **anode** is impure copper and dissolves into the **electrolyte**.
  - The **cathode** is 'plated' with new copper.

An electrolysis cell.

B–A*

- **Electrolysis** is the break-up of a chemical compound (the electrolyte) when you use an electric current.

- In the purification of copper, the electrolyte is a solution of copper (II) sulphate which is electrolysed using copper electrodes.
  - Instead of the electrolyte breaking apart, the **anode dissolves** and the **cathode** is plated in **pure copper**.
  - A sheet of pure copper is used as a cathode. This sheet gets thicker as more pure copper is plated onto it.
  - The impure copper anode is called **blister copper**. Sometimes it's also called boulder copper.
  - The impurities from the copper anode sink to the bottom of the cell.

Recycling copper

D–C

- Copper has a fairly **low melting point** that makes it easy to melt down and recycle. However, copper that's been already used may be contaminated with other elements, such as solder. This means that it can't be used for purposes where the copper must be very pure, such as electric wiring.

- Copper used for recycling has to be sorted carefully so that valuable 'pure' copper scrap isn't mixed with less pure scrap.

- When impure copper is used to make **alloys**, it must first be analysed to find out how much of each element is present. Very impure scrap copper has to be electrolysed again before it can be used.

Alloys

D–C

- Most **metals** form **alloys**.
  - Amalgam contains mercury.
  - Solder contains lead and tin.
  - Brass contains copper and zinc.

Top Tip!

Alloys are often more useful than the metals they're made from.

B–A*

- **Smart alloys** can:
  - be bent more than steel so they're much harder to damage
  - change shape at different temperatures, called '**shape memory**'.

- New ways of using smart alloys are being discovered. Here are some uses.
  - In the frames of glasses to stop them breaking.
  - In shower heads to reduce the water supply if the temperature gets so hot that it scalds.
  - Surgeons can put a small piece of metal into a person's blocked artery and then warm it slightly. As it warms up, it changes shape into a much larger tube that holds the artery open and reduces the risk of a heart attack.

- **Nitinol** is a smart alloy made from nickel and titanium.

Questions

Grades D–C

1 Which electrode becomes pure copper in electrolysis?

Grades B–A*

2 In the purification of copper, what's the electrolyte?

Grades D–C

3 What metals form the alloy brass?

Grades B–A*

4 Which property do smart alloys have that depends on temperature?

Cars for scrap

Rusting and corrosion

D–C

- **Acid rain** and **salt water** accelerate **rusting**. In winter, icy roads are treated with salt, which means that car bodies rust quicker.

- **Aluminium** doesn't corrode in moist air because it has a protective layer of **aluminium oxide** which doesn't flake off the surface.

B–A*

- Rusting is a **chemical reaction** between iron, oxygen and water called **oxidation**. This is because iron reacts with oxygen to make an oxide.

- The chemical name for rust is **hydrated iron(III) oxide**.

- The word equation for rusting is:

  iron + water + oxygen ⟶ hydrated iron(III) oxide

Materials used in cars

D–C

- **Alloys** often have different and more useful properties than the pure metals they're made from. **Steel** (an alloy made of **iron** and **carbon**) is stronger and harder than iron and doesn't rust as easily as pure iron.

- Steel and aluminium can both be used to make **car bodies**, but each material has its advantages.

steel	aluminum
stronger and harder than aluminium which is important in the event of a crash	mass of a car body is less than the same car body made from steel, so has a better **fuel economy**
car body is cheaper than one made of aluminium	aluminium car body will corrode less so the car body has a much **longer lifetime**

Top Tip!

For grades D-C and B-A*, you need to know the information in this table.

B–A*

Recycling

D–C

- More recycling of **metals** means that less metal ore needs to be mined.

- Recycling of **iron** and **aluminium** saves money and energy compared to making them from their ores.

- Recycling **plastics** means less crude oil is used and less non-biodegradeable waste is dumped.

- Recycling of **glass** has been happening for many years very successfully.

- Recycling **batteries** reduces the dumping of toxic materials into the environment.

- European Union law requires 85% of a car to be recyclable. This percentage will increase to 95% in the future. Technology has to be developed to separate all the different materials used in making a car.

Questions

Grade D-C

1 Explain how aluminium is protected from corrosion in moist air.

Grade B-A*

2 What's the chemical name for rust?

Grade D-C

3 Why is the fuel economy better in an aluminium car rather than one of steel?

Grade B-A*

4 Give one reason why a car should be made from aluminium and one why it should be made of steel.

Clean air

What's in clean air?

D–C

- Clean air is made up of 78% nitrogen, 21% oxygen and of the remaining 1%, only 0.035% is carbon dioxide.

- These percentages don't change very much because there's a balance between processes that use up and make carbon dioxide and oxygen. These processes are shown in the **carbon cycle**.

B–A*

- Over the past few centuries the percentage of carbon dioxide in the air has increased due to:
  – **increased energy usage** – more fossil fuels are being burnt in power stations
  – **increased population** – the world's energy requirements increase
  – **deforestation** – as more rainforests are cut down, less photosynthesis takes place.

The atmosphere

D–C

- Scientists **know** that gases trapped in liquid rock under the surface of the Earth are always escaping. This happens in volcanoes.

- Scientists **guess** about the original atmosphere of the Earth. It's known that microbes developed that could photosynthesise. These organisms could remove carbon dioxide from the atmosphere and add oxygen. Eventually the level of oxygen reached what it is today.

- Scientists **know** that the gases came from the centre of the Earth in a process called **degassing**.

B–A*

- They **think** that the original atmosphere contained ammonia and later carbon dioxide. A chemical reaction between ammonia and rocks produced nitrogen and water. The percentage of nitrogen slowly increased, and as it is unreactive, very little nitrogen was removed. Much later, organisms that could photosynthesise evolved and converted carbon dioxide and water into oxygen. As the percentage of oxygen in the atmosphere increased, the percentage of carbon dioxide decreased, until today's levels were reached.

Pollution control

D–C

- You need to be able to describe the origin of these atmospheric pollutants.

pollutant	carbon monoxide	oxides of nitrogen	sulphur dioxide
origin of pollutant	incomplete combustion of petrol or diesel	formed in the internal combustion engine	sulphur impurities in fossil fuels

B–A*

- A car fitted with a **catalytic converter** changes carbon monoxide into carbon dioxide.

- A reaction between nitric oxide and carbon monoxide takes place on the surface of the catalyst. The two gases formed are natural components of air.

carbon monoxide + nitric oxide ⟶ nitrogen + carbon dioxide

$$2CO + 2NO \longrightarrow N_2 + 2CO_2$$

Questions

(Grades D-C)

1 Which process in the carbon cycle uses up carbon dioxide?

(Grades B-A*)

2 Explain why deforestation causes problems for the atmosphere.

(Grades D-C)

3 What does a catalytic converter do?

(Grades B-A*)

4 Write down the equation for removing carbon monoxide from the atmosphere using a catalytic converter.

Faster or slower (1)

Simple collision theory

- The more collisions there are in a reaction, the faster the reaction.

- The rate of a chemical reaction can be increased by increasing the **concentration** and **temperature**.
  - As the concentration increases the particles become more crowded. This means there are more collisions so the rate of reaction speeds up.
  - The reacting particles have more **kinetic energy** and so the number of collisions increases.

- **Collision frequency**, *not* the number of collisions, determines the rate of a reaction. This describes the number of **successful collisions** between reactant particles each second.
  - For a successful collision, each particle must have lots of kinetic energy.
  - As the concentration increases, the number of collisions per second increases and so the rate of reaction increases.
  - As the temperature increases, the reactant particles have more kinetic energy so there are more energetic collisions. These more energetic collisions are more successful.

Rates of reaction

- Magnesium ribbons of hydrochloride acid were reacted in some experiments.

Graph A shows how rate of reaction changes with a change in **concentration** of the reactants.

Graph B shows how rate of reaction changes with a change in **temperature**.

- The total volume of hydrogen produced during all experiments is the same because excess acid and the same mass of magnesium are used.

- The rate of reaction can be worked out from the **gradient** of a graph, which can be found by drawing construction lines.
  - Choose a part of the graph where there's a straight line (not a curve).
  - Measure the value of y and x.
  - Then divide y by x.
  - The gradient of the graph is: gradient $= \dfrac{y}{x}$

Top Tip!

Extrapolation means extending the graph to read it. **Interpolation** means reading within the graph between closer points.

Questions

Grades D-C

1 Use ideas about particles to explain how the rate of a reaction can be altered.

Grades B-A*

2 Explain what's meant by 'collision frequency'.

Grades D-C

3 If a reaction between the same mass of magnesium and excess acid is measured at two different temperatures, the total volume of gas produced doesn't change. Explain why.

Grades B-A*

4 What does 'extrapolation' mean?

Faster or slower (2)

D–C

Explosions

- During an **explosion**, a large volume of gaseous products are released, moving outwards from the reaction at great speed causing the explosive effect.

- **Combustible powders** often cause explosions.
  - A powder reacts with oxygen to make large volumes of carbon dioxide and water vapour.
  - A factory using combustible powders such as sulphur, flour, custard powder or even wood dust must be very careful. The factory owners must ensure that the powders can't reach the open atmosphere and that the chance of producing a spark is very small.

D–C

Surface area

- A powdered reactant has a much larger **surface area** than the same mass of a block of reactant.

- As the surface area of a solid reactant increases so does the rate of reaction.

- Fewer reacting particles of **B** can be in contact with reacting particles of **A**. As the surface area increases there are more collisions, which means the rate of reaction increases.

small surface area   large surface area

  ● reacting particle of substance **A**
  ● reacting particle of substance **B**

- The graph shows the rate of reaction between calcium carbonate and dilute hydrochloric acid.
  - As the reaction takes place, the mass on the balance decreases. This is because carbon dioxide gas is escaping.

$$CaCO_3 + 2HCl \longrightarrow CaCl_2 + H_2O + CO_2$$

  - The **gradient** of the graph is a measure of the rate of reaction. As the reaction takes place, the rate of reaction becomes less and less because the concentration of acid and the mass of calcium carbonate decrease.
  - As the reaction proceeds there are fewer collisions between reactants.

B–A*

- It's the **collision frequency** between reactant particles that's important in determining how fast a reaction takes place. The more successful collisions there are each second, the faster the reaction.

- When the surface area of a solid reactant is increased, there will be more collisions each second. This means the rate of reaction increases.

D–C

Catalysts

- A **catalyst**:
  - increases the rate of a reaction
  - is unchanged at the end of a reaction
  - is needed in small quantities to catalyse a large mass of reactants.

B–A*

  - usually only makes a **specific** reaction faster
  - doesn't increase the number of collisions per second
  - works by making the collisions more successful
  - helps reacting particles collide with the correct orientation
  - allows collisions between particles with less kinetic energy than normal to be successful.

Questions

Grades D–C

1. Give two examples of combustible powders.
2. Why is only 50 cm³ of hydrogen gas produced when using a 0.135 g lump of zinc with acid, compared to 100 cm³ of gas produced when using a 0.27 g lump of zinc?

Grades B–A*

3. What happens to the number of particle collisions if the surface area of a reactant is increased?
4. How does a catalyst help speed up a reaction?

C2 Summary

Paints and pigments

Paints are made from:
- coloured rock (**pigment**) particles
- **solvent**
- **binding medium**.

Paint is a **colloid**. Oil paints 'dry' by oxidation with atmospheric oxygen.

Paints are applied as thin layers, which dry when the solvent evaporates.

Building materials

These are manufactured from rocks:
- brick is made from clay
- glass is made from sand
- aluminium or iron are made from their ores.

When limestone is heated to thermally decompose it, one substance is changed into two: calcium oxide and carbon dioxide.

Metals

Metals from cars are easy to recycle. Plastics need legislation.

Pure metals can be mixed with other elements to make **alloys**.
- copper and zinc make brass
- lead and tin make solder.

Aluminium doesn't corrode in moist conditions because it has a protective layer of aluminium oxide which doesn't flake on the surface.

Iron and aluminium are used in cars. Iron rusts, but aluminium doesn't. Rusting is the oxidation of iron to form hydrated iron oxide.

Alloys have properties that make them more useful than the pure metal. For example, steel is harder and stronger than iron.

Earth and atmosphere

The Earth is made of **tectonic plates** that float on the mantle. The plates are moving all the time.

Tectonic plates are found on top of the mantle as they're less dense than the mantle.

If molten rock cools slowly the crystals that are formed are bigger.

The atmosphere used to be poisonous, but then plants produced oxygen. The composition of the atmosphere is now 21% oxygen and 78% nitrogen.

Carbon dioxide is given out in **combustion** but taken in during **photosynthesis**.

The Earth's original atmosphere came from gases escaping from the interior of the earth.

Fast or slow?

Rates of reaction are affected by:
- **temperature**
- **surface area**
- **concentration**
- **catalysts**.

An increase in surface area of a reactant increases the frequency of collisions.

The higher the temperature the faster the particles move. This increases the rate of reaction.

Collecting energy from the Sun

Photocells

D–C

- The **advantages** of **photocells** are:
  - they're robust and don't need much maintenance
  - they don't need any fuel or long power cables
  - they don't cause pollution or contribute to climate change
  - they use a renewable energy resource.

- The only **disadvantage** is that they don't produce electricity when it's dark.

B–A*

- A photocell contains two pieces of **silicon** joined together to make a **p-n junction**. An electric field is created between the two pieces.
  - One piece has an impurity added to produce an **excess** of **free electrons** – n-type silicon.
  - The other piece has a different impurity added to produce an absence of free electrons – **p-type silicon**.

How is an electric field set up in a photocell?

- Sunlight contains energy packets called **photons**. Photons cause free electrons to move producing an electric current.

Solar heating

D–C

- A house that uses **passive solar heating** makes use of direct sunlight. It has large windows facing the Sun (South) and small windows facing North.
  - During the **day**, energy from the Sun warms the walls and floors.
  - During the **night**, the walls and floors radiate energy back into the room.

- Curved solar reflectors **focus** energy from the Sun.

Top Tip!

In the Southern Hemisphere, the larger windows will need to face North.

- **Solar reflectors** are moved by a computer to make sure they're always facing the Sun.

B–A*

- The temperature inside a **greenhouse** is higher than on the outside.
  - The Sun is very hot and produces **radiation** with a short wavelength. This radiation passes through glass and is **absorbed** by plants and soil.
  - The plants and soil aren't as hot and re-radiate energy with a longer wavelength (infrared).
  - This radiation isn't able to pass through glass and is **reflected** back into the greenhouse.

Wind energy

D–C

- Moving air has **kinetic energy** which is transferred into electricity by a **wind turbine**.

- **Wind farms** don't contribute to global warming, nor do they pollute the atmosphere.

B–A*

- They don't work if there isn't any wind, or if the wind speed is too great. Furthermore, they can be noisy and impact on the countryside.

Questions

Grades D-C

1 Some roadside warning signs are powered using photocells. Suggest why.

Grades B-A*

2 What's the difference between n-type and p-type silicon?

Grades D-C

3 Why do the large windows in a solar-heated house have to face North in Australia but South in England?

Grades B-A*

4 What type of radiation, from the electromagnetic spectrum, is radiated by plants and soil in a greenhouse?

Generating electricity

Generators

- The **current** from a **dynamo** can be increased by:
  - using a **stronger magnet**
  - **increasing** the number of **turns** on the coil
  - **rotating** the magnet **faster**.

Top Tip!
It's the relative movement of the magnet and coil that's important.

- The voltage and frequency from a dynamo can be displayed on an **oscilloscope**.

- The formula to work out the frequency is:
  frequency (in hertz, Hz) = 1 ÷ period (in seconds, s)

- The **generator** at a **power station** works like a dynamo.

- A simple generator consists of a coil of wire rotating between the poles of a magnet to produce a current in the coil.

A simple alternating current (ac) generator.

- In a power station, **fuel** is burned to heat water to produce **steam**. Steam at high pressure turns a **turbine** which then drives a generator.

D–C

Distributing electricity

- The **National Grid** distributes electricity around the country at **high voltages**. This means that:
  - there's less energy loss
  - the distribution costs are lower
  - electricity prices are cheaper.

- Some **transformers** in the National Grid increase (**step-up**) voltages, others decrease (**step-down**) voltages.

D–C

- Electricity in the UK is generated with a frequency of 50 Hz.

- When a current passes through a wire, the wire gets hotter. The greater the current, the hotter the wire becomes.

- When a transformer **increases** the **voltage**, the **current decreases**. There's less heating effect and less energy is lost to the environment.

- The formula to work out **energy efficiency** is: $\text{efficiency} = \dfrac{\text{electrical energy output}}{\text{fuel energy input}}$

- Energy in a power station is lost in the boilers, cooling towers and generator.
  fuel energy input = waste energy output + electrical energy output

> For every 1000 J of energy stored in coal, 150 J is wasted in the boiler, 450 J is wasted in the cooling towers and 70 J in the generator.
> electrical energy output = 1000 – (150 + 450 + 70) = 330 J
>
> $$\text{efficiency} = \frac{330}{1000} = 0.33 \text{ or } 33\%$$

B–A*

Questions

Grades B-A*

3 Explain why electricity is distributed around the country at high voltages.

Grades D-C

1 Suggest why steam is used to turn a turbine at most power stations.

2 What are the advantages of distributing electricity at high voltages?

4 For every 1000 J of energy stored in coal, 160 J are lost in a power station boiler. A further 500 J are lost in the cooling towers and 40 J in the generator. Calculate the efficiency of the power station.

Fuels for power

Fuels

D–C

- A **fuel** burns in air to release energy in the form of heat.

- **Biomass** can be burned. It can be **fermented** to produce **methane**, which is then burned.

- **Nuclear fuels** don't burn. In a nuclear reactor, uranium atoms split and release lots of energy as heat. Nuclear reactions in power stations have to be controlled to avoid an explosion.

B–A*

- **Fossil fuels** are **non-renewable** and produce **carbon dioxide**, a greenhouse gas that contributes to climate change.

- **Biomass** energy sources are **renewable**, but still produce carbon dioxide when used as a fuel.

Power

D–C

- **Power** is a measure of the rate at which energy is used. The unit is the **watt** (W).
  power = voltage × current

- **Electrical consumption** is the amount of energy that's been used. The unit is the **kilowatt hour** (kWh). energy used = power × time

- The formula to work out the **cost** of using an electrical appliance is:
  cost of electricity used = energy used × cost per kWh

B–A*

- Electricity is cheaper at night when not as much is needed but it still has to be produced.

Nuclear power

D–C

- **Radiation** from nuclear waste causes **ionisation**, which causes a change in the structure of any atom exposed to the radiation.

- The cells of our bodies are made of many different atoms. DNA is an important chemical in a cell and can be changed when it's exposed to radiation. The cell behaves differently to normal and this is called **mutation**. One effect of mutation is for a cell to divide in an uncontrolled way. This can lead to **cancer**.

- **Waste** from a nuclear reactor can remain radioactive for thousands of years.

- **Plutonium**, one of the waste products, can be used to make nuclear weapons. Nuclear bombs destroy everything in a very large area and make that area unusable for a long time.

B–A*

- Here are some **advantages** of using **nuclear power**:
  – fossil fuel reserves aren't being used
  – there's no atmospheric pollution
  – no carbon dioxide is produced so no global warming effects.

- And some **disadvantages**:
  – nuclear power stations have high maintenance costs, decommissioning costs and produce quantities of radioactive waste
  – there's a risk of a major accident similar to the one at Chernobyl.

Questions

Grades D-C

1 How much does it cost to use a 2 kW toaster for $\frac{1}{4}$ hour if electricity costs 10p per kWh?

Grades B-A*

2 Suggest one household appliance that could use cheaper electricity at night.

Grades D-C

3 Explain how exposure to nuclear radiation can cause cancer.

Grades B-A*

4 Describe the advantages of producing electricity from nuclear power stations.

Nuclear radiations

Nuclear radiation

- Most **background radiation** is naturally occurring.

- Alpha, beta and gamma radiations come from the **nucleus** of an atom.

- **Alpha** radiation causes **most** ionisation and **gamma** radiation the **least**. Ionisation produces **charged particles**.

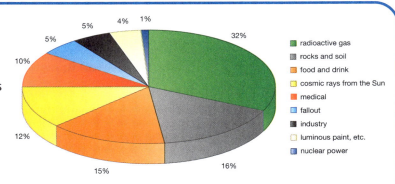

- 32% radioactive gas
- 16% rocks and soil
- 15% food and drink
- 12% cosmic rays from the Sun
- 10% medical
- 5% fallout
- 5% industry
- 4% luminous paint, etc.
- 1% nuclear power

Sources of background radiation. Which source contributes most to the background radiation?

type of ionising radiation	range	penetration through materials
alpha	short – a few centimetres	easily absorbed by a sheet of paper or card
beta	about 1 metre	absorbed by a few millimetres of aluminium
gamma	theoretically infinite	can pass through metres of lead or concrete

- **Atoms** contain the same number of **protons** and **electrons** – this means they're neutral.

- **Ionisation** involves gaining or losing electrons:
  – when the atom **gains** electrons, it becomes **negatively charged**
  – when the atom **loses** electrons, it becomes **positively charged**.

D–C

B–A*

Uses of radioactivity

- **Alpha radiation** is used in **smoke alarms**.

- **Beta** or **gamma** sources are used in **rolling mills** to conrol **thickness**.

- **Gamma** sources are injected into the body as **tracers**.

D–C

Radioactive waste

- Some **radioactive waste** can be reprocessed into new, useful radioactive material.

- Radioactive waste must be **stored safely** because it can remain radioactive for thousands of years. It must be stored where it can't leak into natural underground water supplies and then into lakes and rivers.

- Radioactive waste isn't suitable for making nuclear bombs, but it could be used by terrorists to contaminate water supplies or areas of land.

D–C

B–A*

Questions

Grades D-C

1 What fraction of background radiation comes from natural sources?

Grades B-A*

2 Why does an atom become positively charged when it loses electrons?

Grades D-C

3 A rolling mill uses beta radiation to control the thickness of cardboard. Explain why alpha radiation and gamma radiation aren't used.

Grades B-A*

4 Suggest why radioactive waste shouldn't be stored underground near to a tectonic plate boundary.

Our magnetic field

Magnetic fields

D–C

- The **magnetic field** around a coil of wire is similar to the field around a bar magnet.

- A magnet shouldn't be brought close to a **television** or **computer monitor** because it causes the electron beam in the tube to change direction. The beam strikes the wrong part of the screen giving a distorted picture.

B–A*

- When a coil of wire moves inside a magnet, a current is produced in the coil. This is the principle of the **dynamo**.

- Within the core of the Earth, the solid inner core moves within the liquid outer core. Electric currents are produced that create the Earth's magnetic field.

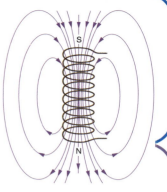

The magnetic field around a coil of wire carrying a current.

Origin of the Moon

D–C

- When the Solar System was formed, there were probably more planets than today.

- Scientists believe there was another planet in the same orbit as the Earth.
  – The two planets collided and were both almost totally destroyed.
  – Iron became concentrated at the core of the new Earth, less dense rocks started to orbit.
  – These rocks clumped together to form the Moon.
  – The Earth and the Moon were rotating much faster than they do now – the Moon caused the speed of rotation to slow down.

B–A*

- The core of the Moon doesn't contain iron. This suggests that the debris came from the **mantle**. The iron from the core of the other planet fused with the iron from the core of 'old' Earth.

- The average **density** of the Earth is nearly twice that of the Moon.

- The Moon has the same **oxygen composition** as the Earth - other planets and meteorites don't.

Cosmic rays

B–A*

- **Cosmic rays** are streams of charged particles originating from the Sun and other bodies in space.
  – They travel very quickly and have lots of energy.
  – They interact with the Earth's magnetic field as they spiral around field lines concentrating at the poles.
  – They're the cause of **auroras** – patterns of light seen in the sky at high latitudes.

Solar effects

D–C

- The **energy** from a **solar flare** is equivalent to that from a million hydrogen bombs.

- Large numbers of charged particles are emitted at very high speeds. These produce magnetic fields that interact with the Earth's magnetic field.

B–A*

- Although only a small fraction of the energy from solar flares reaches the Earth, it's enough to interfere with radio and microwave signals and affect power distribution through the National Grid.

Questions

Grades D-C

1 A television screen is coated with red, blue and green dots. What happens to the dots when they are hit by a beam of electrons?

2 How do scientists believe the Moon was formed?

Grades B-A*

3 What evidence supports the idea that the Moon was formed at the same time as the Earth?

4 Describe how the Aurora Borealis (Northern Lights) are formed.

Exploring our Solar System

The Universe

D–C

- The Earth is one of the planets in our Solar System orbiting the Sun.

- In August 2006, scientists decided that Pluto shouldn't be considered a planet because of its size and orbit shape.

1. Mercury
2. Venus
3. Earth
4. Mars
5. Jupiter
6. Saturn
7. Uranus
8. Neptune
9. Pluto

Top Tip!
Use a mnemonic to remember the order of the planets: **M**any **V**ets **E**arn **M**oney **J**ust **S**itting **U**nder **N**uts (**P**lants).

B–A*

- **Comets** have **elliptical orbits**. They pass inside the orbit of Mercury and out beyond the orbit of Pluto.

- A **meteor** is made from grains of dust that burn up as they pass through the Earth's atmosphere. They heat the air around them which glows, and the streak of light seen is called a 'shooting star'.

- **Black holes** are formed where large stars used to be. You can't see a black hole because light can't escape from it.

- **Moons** orbit planets and planets orbit stars because **centripetal force** acts on them.
  – Centripetal force acts towards the **centre** of the circular orbit.
  – **Gravitational attraction** is the source of the centripetal force.

Which planet is closest to the Sun?
Centripetal force acts towards the centre of the Sun.

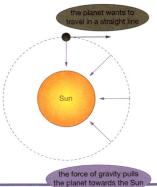

the planet wants to travel in a straight line

Sun

the force of gravity pulls the planet towards the Sun

Exploration

D–C

- Unmanned spacecraft (**probes**) have explored the surface of the Moon and Mars. They can go to places where humans can't survive.

- The **Hubble Space Telescope** orbits the Earth collecting information from distant galaxies.

- The Moon is the only body in space visited by humans. **Astronauts** wear normal clothing in pressurised spacecraft, but outside they need to wear special **spacesuits**.
  – A dark visor stops an astronaut being blinded.
  – The suit is pressurised and has an oxygen supply for breathing.
  – The surface of the suit facing towards the Sun can reach 120 °C.
  – The surface of the suit facing away from the Sun may be as cold as –160 °C.

- When travelling in space, astronauts experience **lower gravitational forces** than on the Earth.

Light-years

B–A*

- Distances in space are very large and are measured in **light-years**.

- Light travels at 300 000 km/s.
  – Light from the Sun takes about 8 minutes to reach us.
  – Light from the next nearest star (Proxima Centauri) takes 4.22 years.

Top Tip!
A light-year is a unit of distance: the distance light travels in one year.

Questions

Grades D-C

1 Write down the names of the planets in the Solar System in order from the Sun.

Grades B-A*

2 A planet, P, is in circular orbit about a star, S. What happens to the motion of the planet if the star suddenly stops exerting a gravitational force on the planet?

Grades D-C

3 Why do astronauts have to wear special spacesuits when outside their spacecraft?

Grades B-A*

4 Calculate the approximate distance of the Earth from **a** the Sun **b** Proxima Centauri.

Threats to Earth

Asteroids

D–C

- Asteroids are **mini-planets** or **planetoids** orbiting the Sun in a 'belt' between Mars and Jupiter. They're large rocks that were left over from the formation of the Solar System.

- **Geologists** have examined evidence to support the theory that asteroids have collided with the Earth.
  - Near to a crater thought to have resulted from an asteroid impact, they found quantities of the metal **iridium** – a metal not normally found in the Earth's crust but common in meteorites.
  - Many fossils are found below the layer of iridium, but few fossils are found above it.
  - **Tsunamis** have disturbed the fossil layers, carrying some fossil fragments up to 300 km inland.

B–A*

- All bodies in space, including planets, were formed when clouds of gas and dust collapsed together due to **gravitational forces of attraction**.

- The mass of an object determines its gravitational force. **Asteroids** have relatively low masses compared to the mass of **Jupiter**. Jupiter's gravitational force prevents asteroids from joining together to form another planet, so they remain in a 'belt' as they orbit the Sun.

- Sixty-five million years ago, an asteroid with a diameter of 10 km struck the Earth. Dust from the collision passed into the atmosphere, blocking sunlight and lowering temperatures. Photosynthesis couldn't take place and herbivorous dinosaurs couldn't feed. This ultimately led to the extinction of all dinosaurs.

Comets

D–C

- Compared to the near circular orbits of the planets, the orbit of a **comet** is very **elliptical**.
  Most comets pass inside the orbit of Mercury and well beyond the orbit of Pluto.

- As the comet passes close to the Sun, the ice melts. Solar winds blow the dust into the **comet's tail** so the tail always points away from the Sun.

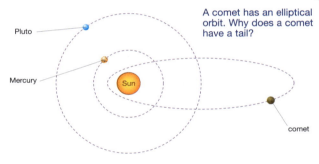

A comet has an elliptical orbit. Why does a comet have a tail?

Pluto

Mercury

Sun

comet

B–A*

- The **speed** of a **comet** increases as it approaches the Sun and decreases as it gets further away. This is a result of changing gravitational forces.

NEOs

D–C

- Scientists are constantly monitoring and plotting the paths of comets and other near-Earth-objects (**NEOs**).

B–A*

- If a NEO was on a collision course with the Earth, one option would be to launch a rocket filled with explosives. This could alter its course enough to miss the Earth.

- If a NEO did hit the Earth, the result could be the end of life on the Earth as we know it.

> **Top Tip!**
> The longer we study and plot the path of a NEO, the more accurate our prediction about its future movement and the risk of collision will be.

Questions

Grades D-C

1 What's an asteroid?

Grades B-A*

2 Explain why carnivorous dinosaurs became extinct sixty-five million years ago.

Grades D-C

3 Why is it important to constantly monitor the paths of comets and NEOs?

Grades B-A*

4 How could a collision between a NEO and the Earth be averted?

The Big Bang

The Universe

- Almost all of the **galaxies** in the Universe are moving away from each other. The furthest galaxies are moving fastest. The Universe is expanding all the time.

- **Microwave signals** are constantly reaching Earth from all parts of the Universe.

- When a source of light is moving away from an observer, its **wavelength** appears to increase. This shifts light towards the red end of the **visible spectrum – red shift**.

- When scientists look at light from the Sun, there's a pattern of lines across the spectrum. This pattern moves towards the red end of the spectrum when they look at light from distant stars. The faster a star is travelling, the greater the red shift.

a The spectrum of white light from the Sun (the black lines show that helium is present) and **b** the spectrum of light from a distant star showing red shift.

- Scientists can use information from red shift to work out the age of the Universe.

D–C

B–A*

Stars

- A medium-sized star, like the Sun, becomes a **red giant**: the core contracts, the outer part expands and cools, and it changes colour from yellow to red. During this phase, gas shells, called **planetary nebula**, are thrown out.

- The core becomes a **white dwarf** shining very brightly, but eventually cools to become a **black dwarf**.

- Large stars become **red supergiants** as the core contracts and the outer part expands. The core suddenly collapses to form a **neutron star** and there's an explosion called a **supernova**. Remnants from a supernova can merge to form a **new star**. The dense core of the neutron star continues to collapse until it becomes so dense it forms a **black hole**.

- The swirling cloud of gas and dust is a **nebula**. Nebula clouds are pulled together by **gravity** into a spinning ball of gas, which starts to get hot and glow.

- This **protostar** is shining but can't be seen because of the dust cloud. Gravity causes the star to become smaller, hotter and brighter.

- After millions of years, the core temperature is hot enough for **nuclear fusion** to take place. As hydrogen nuclei join together to form helium nuclei, energy is released. The star continues to shine while there's enough hydrogen.

- Small stars shine for longer than large stars because although they have less hydrogen, they use it up at a slower rate.

- What happens at the end of a star's life depends on its size.

D–C

B–A*

isotopes of hydrogen → helium + neutron + energy

What has to happen to a protostar before nuclear fusion can take place?

Questions

Grades D-C

1 Which galaxies in the Universe are moving fastest?

Grades B-A*

2 Explain how scientists can calculate the age of the Universe using red shift.

Grades D-C

3 Our Sun is about 4.6 billion years old. Patrick says it may have been a star before that. Explain how this is possible.

Grades B-A*

4 Explain why smaller stars live longer than larger stars.

P2 Summary

Energy sources

Kinetic energy from moving air turns the blades on a wind turbine to produce electricity.

The **Sun** is a stable energy source. It **transfers energy** as light and heat to Earth.

Photocells use the Sun's light to produce electricity.

Passive solar heating uses glass to help keep buildings warm.

Electricity generation

A **dynamo** produces electricity when coils of wire rotate inside a magnetic field. The size of the current depends on:
– the number of turns
– the strength of the field
– the speed of rotation.

– **Transformers** change the size of the voltage and current.
– The National Grid transmits electricity around the country at **high voltage** and **low current**. This reduces energy loss.

Nuclear fuels are radioactive. The radiation produced can cause cancer. Waste products remain radioactive for a long time.

– **Fossil fuels** and **biomass** are burned to produce heat.
– Nuclear fuels release energy as heat.
– Water is heated to produce steam.
– the steam drive turbines
– turbines turn generators
– generators produce electricity.

The main forms of **ionising radiation** are:
– **alpha**
– **beta**
– **gamma**.
Their uses depend on their penetrative and ionisation properties.

The Earth's field

– The **Earth** is surrounded by a **magnetic field** similar in shape to that of a bar magnet or coil of wire.
– The core of the Earth contains molten iron.

Solar flares are the result of clouds of charged particles being emitted from the Sun at high speed disturbing the magnetic field around Earth.

When two planets collide, a new planet and a moon may be formed.

The Earth and Universe

– **Planets, asteroids** and **comets** orbit the Sun in our **Solar System**.
– The Universe consists of many galaxies.
Centripetal forces keep bodies in orbit

Most **asteroids** are between Mars and Jupiter but some pass closer to Earth. They are constantly being monitored. An asteroid strike could cause climate change and species extinction.

– Scientists believe that the Universe started with a **Big Bang**.
– **Stars** have a **finite life** depending on their size.

The **Universe** is explored by telescopes on Earth and in space. Large distances mean that it takes a long time for information to be received and inter-galactic travel unlikely.

Molecules of life

Cells

D–C

- **Cell respiration** is carried out inside **mitochondria**. During respiration, energy is released from glucose in the presence of oxygen.

DNA

B–A*

- The structure of DNA helps it to copy itself every time a cell divides. This is called **DNA replication**.

- DNA is split into sections called **genes**. Each gene holds the code for making a protein that our bodies need by using amino acids from our food. Proteins are made by joining amino acids into a chain. The DNA controls the order of amino acids and the production of proteins is called **protein synthesis**.

- DNA can be used to identify people by **DNA fingerprinting**, which produces a pattern of unique bands like a barcode:
  - blood or cell sample is isolated
  - DNA is extracted
  - restriction enzymes are used to fragment the DNA
  - DNA fragments are placed on gel
  - fragments are separated by an electric curent (electrophoresis)
  - banding of DNA fingerprint can be matched.

- The order of bases found in DNA is called the base code. Each three bases code for an amino acid. Cells use these base codes to join amino acids together in the correct order.

- Amino acids can be changed in the liver into any that are missing from the diet.

DNA replication. The double strand 'unzips' and two new double strands are formed.

new bases pair up with their complementary base to form two new double strands

complementary base pairing, the cross-links in DNA are formed from four bases called A, T, C, G. Base A always links with base T and base C always links with base G

template for making a new DNA strand

when DNA replicates, the base pairs separate and the strand 'unzips' to form two single strands

DNA double helix

Enzymes

D–C

- An **enzyme** is a biological **catalyst**; it's a protein that speeds up a biological reaction. Enzymes catalyse most chemical reactions occurring within cells such as respiration, photosynthesis and protein synthesis. Each enzyme is **specific** to a **substrate**. In an enzyme-catalysed reaction, substrate molecules are changed into **product** molecules.

B–A*

- Each enzyme has a unique sequence of amino acids so each enzyme has a different shape. Within this shape is an **active site**. Once the substrate is attached to the active site it is turned into a **product**. The enzyme is like a lock and the substrate like a key. Enzymes are **denatured** when the shape of the active site changes.

- Enzymes have an **optimum pH** when the active site and the substrate molecule are a perfect fit. Changing the pH denatures the enzyme.

- As the temperature increases, the molecules gain more energy. More collisions occur and the rate of reaction increases. Above the optimum temperature, the enzyme denatures and the reaction stops. Lowering the temperature again will have no effect, as the enzyme shape can't be changed back.

Questions

Grades D–C

3 What is meant by the term 'enzyme'?

Grades B–A*

1 Describe how proteins are made.
2 How many bases code for one amino acid?

Grades B–A*

4 Explain why high temperature stops an enzyme joining to a substrate.

Diffusion

Diffusion

- **Diffusion** is the movement of a substance from an area of high concentration to an area of low concentration.

- **Molecules** move **randomly** in all directions, but most will move from an area of high concentration. Therefore **diffusion** is the net movement of particles from an area of high **concentration** to an area of low concentration due to the random movement of individual particles.

- The **rate of diffusion** can be increased by: increasing the **surface area**, decreasing the diffusion distance or a greater concentration difference.

Diffusion in humans

- **Alveoli** in the **lungs** have a higher concentration of oxygen than the blood that surrounds them, so the oxygen diffuses into the blood. Carbon dioxide diffuses from the blood into the alveoli. To maintain this **gas exchange**, breathing takes place.

- After **eating** there is a high concentration of digested food molecules in the small intestine, which causes them to diffuse through the cells of the small intestine wall into the blood.

- A **foetus** needs to be supplied with food and oxygen from its mother so that it can develop. The mother's blood and the foetus's blood come close together in the **placenta**. Dissolved food and oxygen pass into the foetus's blood and carbon dioxide and waste products pass out into the mother's blood by **diffusion**.

- Villi in the small intestine and alveoli in the lungs have special adaptations to increase the rate of diffusion.

a
a thin wall for easy diffusion of food into the bood
a good blood supply
long, permeable villi to increase surface area
moist and permeable surface area

b
a good blood supply
a thin cell wall

Diffusion in **a** a villus **b** alveoli.

- To speed up movement across the **placenta** it has a **huge surface area** with a thin wall so substances only have a short distance to diffuse.

- A **synapse** is a gap between two neurones (nerve cells). To carry a signal from one neurone to the next, the synapse releases a transmitter substance which can diffuse across the gap between the two neurones.

Diffusion in plants

- Carbon dioxide diffuses into the leaf through small pores called stomata. During photosynthesis, oxygen levels increase inside the leaf, which causes oxygen to diffuse out of the leaf. **Water** is lost by **diffusion** and **evaporation**.

- To increase the rate of **gas exchange**, the leaf has a **large surface area** with lots of **stomata** on the under-surface.

Questions

Grades D-C
1 What is diffusion?

Grades B-A*
2 Explain why villi are important for food absorption.

3 Describe the function of a transmitter substance.

Grades D-C
4 Describe how oxygen is lost from the leaf.

Keep it moving

Blood cells

- **Red blood cells** are adapted to carry as much oxygen as possible. They contain **haemoglobin**, which joins to oxygen, and have **no nucleus** so there's more room to store oxygen. They are **disc-shaped** and have a **dent** on both sides to help absorb oxygen. They are **tiny** so they can carry oxygen to all parts of the body.

- **White blood cells** change shape so they can wrap around (**engulf**) microbes. They don't fight or kill microbes.

- Food, water, hormones, antibodies and waste products are carried in the **plasma**.

- The shape of a red blood cell means it has a large surface area compared to its volume. This enables it to absorb a lot of oxygen.

- In the lungs **haemoglobin** reacts with **oxygen** to form **oxyhaemoglobin**. When it reaches tissue it separates into haemoglobin and oxygen. The oxygen diffuses into the tissue cells and the red blood cells return to the lungs to pick up more oxygen.

Heart

- The **valves** of the heart prevent backflow of blood.

- Coronary arteries supply the heart with **food** and **oxygen**.

- Too much cholesterol in the diet can cause serious **heart problems** and even the need for a **heart transplant**. There are some problems with heart transplants such as: shortage of donors, rejection, dependency on drugs, and difficulty in matching tissue to age and size. Using **mechanical replacements** also has problems, such as the size of the replacement, the power supply needed and the chance of rejection.

RIGHT

LEFT

pulmonary artery, takes deoxygenated blood to the lungs

aorta, takes oxygenated blood to the body

semi-lunar valve

pulmonary vein, brings oxygenated blood from the lungs

vena cava, brings deoxygenated blood from the body

right atrium

left atrium

tricuspid valve

bicuspid valve

valve tendon

right ventricle, thinner wall as pumps blood a relatively short distance to the lungs

left ventricle, has thick muscular wall to pump blood at higher pressure all the way round the body

The structure and function of the heart.

The circulatory system

- **Arteries** transport blood away from the heart. They have thick muscular and elastic walls to help them withstand high blood pressure as blood leaves the heart.

- **Veins** transport blood to the heart. They have a large lumen to help blood flow at low pressure; valves stop blood from flowing the wrong way.

- **Capillaries** join arteries to veins. They have a thin, permeable wall to allow exchange of materials such as oxygen between the capillaries and the body tissue.

- Humans have a **double circulatory system**. One circuit links the heart and lungs, and one circuit links the heart and the body. Blood going to the body can be pumped at a much **higher pressure** than blood going to the lungs. This provides a greater rate of flow to the body.

- Sometimes **cholesterol** in the blood sticks to the inside of artery walls. As it builds up, it forms a plaque that restricts the flow of blood. Some of this cholesterol can break away and block the artery completely.

Questions

Grades D–C

1 Name the chemical in blood that joins to oxygen.
2 Name the blood vessel that takes blood to the lungs.
3 Explain why an artery needs a thick muscular wall.

Grades B–A*

4 What is the advantage of the double circulatory system?

Divide and rule

Multi-cellular

- Humans are made up of millions of cells: they are **multi-cellular**. This gives them many advantages.
  - Multi-cellular organisms can grow large.
  - Cell **differentiation** takes place. Cells change shape or size to carry out different jobs.
  - Organisms become more complex and develop different organ systems.

- The size that a single cell can grow to is limited by its surface area to volume ratio.
  If the cell is too large it can't absorb enough food and oxygen through the surface of its membrane to stay alive. This is why large multi-cellular organisms have developed transport systems.

Mitosis

- Humans have **23 pairs** of **chromosomes**. The chromosomes in a pair look the same and carry similar information. They are called **homologous** pairs. When a cell has pairs of chromosomes it's called a **diploid** cell (i.e. it has the full set of chromosomes).

- During growth, a type of cell division called **mitosis** makes new cells. The new cells are exact copies and contain 23 pairs of chromosomes.

- Mitosis makes **genetically identical** cells.

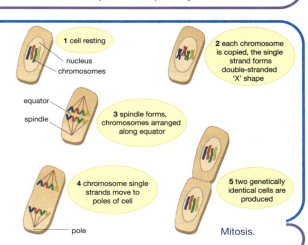

1 cell resting
nucleus
chromosomes

2 each chromosome is copied, the single strand forms double-stranded 'X' shape

equator
spindle

3 spindle forms, chromosomes arranged along equator

4 chromosome single strands move to poles of cell

5 two genetically identical cells are produced

pole

Mitosis.

Meiosis

- Gametes have half a set of chromosomes called the **haploid number**. During fertilisation the gametes join to form a **zygote**. The zygote is diploid and can develop into an embryo.

- **Meiosis** is a special type of cell division that produces gametes. Reproduction using meiosis results in a lot of genetic **variation** within a species.

- Gametes are made when diploid cells divide by meiosis to produce haploid cells. Meiosis involves two divisions: first the pairs of chromosomes separate, then the chromosomes divide in the same way as in mitosis.

pole

1 homologous chromosomes pair up

2 one from each pair moves to opposite poles

3 strands of chromosomes move to opposite poles

4 four new haploid cells form

Meiosis.

Moving sperm

- **Sperm** are specially adapted to swim a long way. Each sperm has lots of **mitochondria** to release energy for motion. **Acrosome** on the sperm head releases enzymes that digest the cell membrane of an egg allowing the sperm inside.

Questions

Grades B-A*

1 Explain why the size of a cell is limited by its surface area to volume ratio.

Grades D-C

2 How many chromosomes are there in a diploid human cell?

Grades D-C

3 Name the type of cell division that produces haploid cells.

Grades B-A*

4 Describe one way in which meiosis and mitosis are the same.

Growing up

Cells

- This table shows how plant and animal cells are similar and different.

plant cell	animal cell
has nucleus, cytoplasm and cell membrane	has nucleus, cytoplasm and cell membrane
cellulose cell wall for support	no cell wall
most have chloroplasts for photosynthesis	no chloroplasts
large vacuole containing cell sap	may have a small vacuole but no cell sap

- A few days after an egg is fertilised it contains a group of cells called **stem cells** which all have the same simple cell structure. They divide and then differentiate to form all the different specialised cells in the body. As the embryo grows all the specialised cells form **tissues** and **organs**.

- Scientists have found ways of making stem cells develop into other specialised cells in the hope of replacing damaged cells. However, many people object to stem-cell research because it can involve human embryos. Scientists use embryos because they are easier to grow than adult stem cells.

Growth

- Animals and plants grow in different ways. Animals tend to grow to a certain size and then stop. Plants can continue to grow.

- The cells of animals and plants cause them to grow in different ways.

plant	animal
most growth is due to cells elongating (growing longer), not dividing	growth is due to cells dividing
cell division only normally occurs at the tips of shoots and roots	cell division occurs all over the body
many cells never lose the ability to differentiate	most animal cells lose the ability to differentiate very early on

Gestation

- **Gestation** is the length of time from fertilisation to birth. The larger the animal the longer gestation tends to be. This is because the animal needs time to develop enough to survive outside the uterus. An elephant has a gestation of 700 days but a rat only has 22 days.

- Different parts of the **foetus** and **baby** grow at different rates. The brain and head develop quickly to co-ordinate the complex human structure and chemical activity.

- After a baby is born, it has regular growth checks. The baby's weight and head size are recorded to check that the baby is growing at a normal rate.

  – Poor weight gain can indicate problems with a baby's digestive system.
  – Larger than normal head size can indicate that fluid is collecting on the brain or that the separate skull bones are not fusing together.

Questions

Grades D-C

1 Name two structures found in plant cells but *not* animal cells.

Grades B-A*

2 Explain why scientists use embryo stem cells instead of adult stem cells.

Grades D-C

3 Explain why elephants have a longer gestation period than mice.

Grades B-A*

4 Suggest a reason for poor weight gain in babies.

Controlling plant growth

Plant hormones

- Farmers, gardeners and fruit growers mostly use man-made **plant hormones** such as **synthetic auxin**. This is sprayed on selected crops to kill weeds and is known as a **selective weedkiller**.

- **Rooting powder** is used to stimulate roots to grow from plant cuttings.

- Hormones are used to make fruit grow without the flowers being fertilised. This means the fruits have no pips, such as seedless grapes.

- A hormone called **ethene** is sprayed on bananas to ripen them ready for sale.

- The seeds taken from a parent plant are **dormant**, which means germination won't take place. Hormones are used to break the dormancy and make the seed germinate.

Responses

- A plant is **sensitive** and responds to different **stimuli**.

- A hormone called **auxin** controls the response. This is made in the tips of roots and shoots and travels through a plant in **solution**.
  - Plant **shoots** grow **towards light** – **positive phototropism**.
  - Plant **roots** grow **away from light** – **negative phototropism**.
  - Plant **shoots** grow **away from** the pull of **gravity** – **negative geotropism**.
  - Plant **roots** grow **with** the pull of gravity – **positive geotropism**.

Top Tip!

The 'positive' and 'negative' terms are difficult to remember. If it's positive it grows towards the stimulus. Imagine being positively attracted to someone!

Shoots are positively phototropic and negatively geotropic.
Roots are negatively phototropic and positively geotropic.

How auxin works

- When the tip of a shoot is cut off it stops growing because it removes the source of **auxin**. If the tip is replaced on the stem it starts to grow again.
  - Auxin is made in the tip.
  - Auxin moves away from light and collects on the shady side of a shoot.
  - Auxin causes cells on the shady side to **elongate** (grow longer) more than cells on the light side.
  - The shady side becomes longer and causes the shoot to bend.

Can you explain why the shoot with substance X is bent?

light all around

untreated seedling

substance **X**

treated seedling

seedlings at start   after 2 days

Questions

Grades D–C

1 Describe the effect of rooting powder on plant cuttings.

2 Which part of the plant shows positive geotropism?

Grades B–A*

3 Describe the effect auxin has on cells in the shoot.

4 Explain why auxin causes the shoot to bend towards the light.

New genes for old

Selective breeding

D–C

- **Selective breeding** is used to breed a cow that has a high yield of creamy milk.
  - Choose types of cows that produce lots of milk (Friesians) or creamy milk (Jerseys).
  - **Cross-breed** them by mating a cow with a bull.
  - Select the best offspring that produce large quantities of creamy milk.
  - Repeat the selection and breeding process for a number of generations.

- Selective breeding often involves animals that are closely related. This is called **inbreeding** and causes a reduction in the **gene pool** (the different genes available in a species). With a smaller gene pool there is less **variation** and the more chance there is of harmful recessive genes being expressed.

B–A*

Mutation

D–C

- Changes to genes, called **mutations**, usually cause harm to the organism. For example, in **haemophilia** the blood doesn't clot.

- Some mutations can be advantageous and give a better chance of survival, such as bacteria that can mutate and become resistant to antibiotics.

- Mutations can be caused by radiation (X-rays), chemicals in cigarette smoke or by chance.

- When a gene mutates the DNA **base sequence** is changed, which alters the protein or even prevents its production.

B–A*

Genetic engineering

D–C

- **Genetic engineering** involves adding a gene to the DNA of an organism. For example, a bacterium has been genetically engineered to make human insulin for people with diabetes.

- Rice doesn't contain vitamin A. Scientists have taken the gene to make **beta-carotene** from carrots and put it into rice plants. Humans eating this rice can then convert the beta-carotene into vitamin A.

rice does not contain vitamin A

carrots do contain vitamin A

gene removed

vitamin A gene

gene inserted into rice

rice now produces vitamin A

Genetically modified rice.

- Scientists are developing crops that are resistant to herbicides, frost and disease. This will enable more crops to grow in difficult places but the gene may have harmful effects on humans who eat the plants.

- Modifying DNA by genetic engineering follows the following basic steps:
  - select the **characteristic**
  - identify and **isolate** the gene
  - **insert** the gene into the chromosome of a different organism
  - **replicate** (copy) the gene in the organism and produce the protein.

Top Tip!

You must be able to debate the issues involved with GM organisms.

B–A*

Questions

Grades D-C

1 Describe the main processes in selective breeding.

Grades B-A*

2 Describe the effect of inbreeding on animals.

Grades D-C

3 Suggest one disadvantage of genetically modified crops.

Grades B-A*

4 List the basic steps in genetic engineering.

More of the same

Cloning animals

D–C

- **Embryo transplantation** can be used to clone cows. Embryo calves are placed in surrogate mothers to develop in the normal way.

- **Human embryos** could also be cloned to provide **stem cells**. These could be transplanted into people suffering from diabetes so that they could make their own insulin. However, people are concerned that this would be unethical because the embryo is a living thing. Some people are also afraid that scientists will clone adult humans.

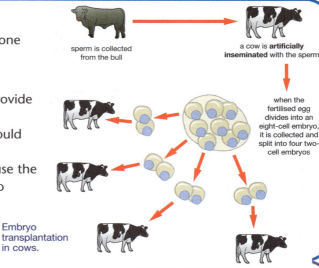

sperm is collected from the bull

a cow is **artificially inseminated** with the sperm

when the fertilised egg divides into an eight-cell embryo, it is collected and split into four two-cell embryos

Embryo transplantation in cows.

each embryo is then **implanted** into a surrogate cow where it grows into a calf. All the calves will be genetically identical to each other, but not to their parents

B–A*

egg cell taken from sheep A and nucleus removed

cells taken from the udder of sheep B and the nucleus removed

nucleus from sheep B is put into egg of sheep A

egg cell is put into a female sheep to grow

Dolly the sheep was cloned using a process called **nuclear transfer**.

cell grows into a clone of sheep B

- There are considerable **risks** involved in cloning:
  – there is a low rate of success
  – research into human cloning raises many moral and ethical issues
  – Dolly died of conditions linked to old age, yet she was only seven years old.

- But there are also some **benefits** of cloning:
  – cloned pigs could make up for a shortage in transplant organs
  – diseases could be cured using embryonic stem cells.

Cloning plants

D–C

- Here are some **advantages** of cloning plants:
  – cloning produces lots of identical plants more quickly
  – cloning enables growers to produce plants that are difficult to grow from seed.

- And some **disadvantages**:
  – the plants are all genetically identical so if the environment changes or a new disease breaks out, it's unlikely that any of the plants will survive
  – cloning plants over many years has resulted in little genetic variation.

B–A*

- Small sections of plant tissue can be cloned using **tissue culture**. This must be carried out using **aseptic technique** (everything has to be sterile).
  – Plants with the desired **characteristics** are chosen.
  – A large number of small pieces of **tissue** are taken from the parent plant.
  – They are put into sterile test tubes that contain **growth medium**.
  – The tissue pieces are left in **suitable conditions** to grow into plants.

- Plants are easier to clone than animals. Many plant cells retain the ability to **differentiate** into different cells but most animal cells don't.

Questions

Grades D-C
1 Name the process used to clone cows.

Grades B-A*
2 Name the process used to clone Dolly the sheep.

Grades D-C
3 Suggest one disadvantage of cloning plants.

Grades B-A*
4 Describe the stages involved in tissue culture.

B3 Summary

Molecules of life

Chromosomes are made of **DNA**. A section of DNA is called a **gene**. Each gene codes for a particular **protein**.

Animals and plant cells have the following parts:
– **membrane**
– **nucleus**
– **cytoplasm**
– **mitochondria**.
Plant cells also have:
– **cell wall**
– **vacuole**
– **chloroplasts**.

DNA is found in the **nucleus** of the cell. DNA carries **coded information**; individuals have their own unique DNA.

DNA fingerprints can be used to identify individuals.

Enzymes are **biological catalysts**, they catalyse chemical reactions in the body.

Proteins are made up of chains of **amino acids**. When the cell makes a new protein it has to join the amino acids together in the correct order. The base sequence in DNA determines the order of amino acids in the protein.

Blood and diffusion

Carbon dioxide and **oxygen diffuse** in and out of plants through the leaves.

Diffusion of substances takes place in the **placenta**. Food and oxygen diffuses into the foetal blood. Carbon dioxide and waste diffuses into the mother's blood.

Blood is moved around the body in:
– **arteries**
– **veins**
– **capillaries**.
The heart pumps the blood to the lungs and body.

In the lungs **oxygen** diffuses into **red blood cells**. **Carbon dioxide** diffuses from the blood into the **lungs**.

Food diffuses from the **small intestine** into the **blood**. The **alveoli** are adapted to increase the rate of diffusion.

Cell division and growth

There are five stages in human growth:
– **infancy**
– **childhood**
– **adolescence**
– **maturity**
– **old age**.

Eggs and sperm are special cells called **gametes**. A type of cell division called **meiosis** produces them. Gametes contain the **haploid** number of chromosomes. **Fertilisation** takes place when an egg and sperm join. The result is a diploid cell called a **zygote**.

Cells divide by **mitosis** so that organisms can grow and replace old cells.

After cells divide they become specialised. This is called **differentiation**. Undifferentiated cells are called **stem cells**. Stem cells can develop into different types of cells, tissues and organs.

Plant hormones

Plant **hormones** have commercial uses such as weedkillers and rooting powder.

The hormone called **auxin** is involved in the plant's response to light (**phototropism**) and the plant's response to gravity (**geotropism**).

Using cells and DNA

Selective breeding involves breeding organisms with the best characteristics.

Cloning involves producing genetically identical copies.

Genetic engineering involves taking **genes** from one organism and putting them into another.

What are atoms like?

Atoms

D–C

- The nucleus of an atom is made up of **protons** and **neutrons**.
- The **atomic number** is the number of protons in an atom.
- The **mass number** is the total number of protons and neutrons in an atom.

	relative charge	relative mass
electron	−1	0.0005 (zero)
proton	+1	1
neutron	0	1

Helium has two protons (red) and two neutrons (green). Helium has a mass number of 4.

A proton has a positive charge. An electron has a negative charge. This atom is neutral because it has the same number of protons and electrons. (Neutrons not shown).

B–A*

- An atom is **neutral** because it has an equal number of electrons and protons. The positive charges balance out the negative charges.
- If a particle has an atomic number of 11, a mass number of 23 and a neutral charge, it must be a sodium atom.

atomic number	11
mass number	23
charge	0
	sodium atom

- An element has the symbol $^{14}_{6}C$, it has no charge so it's an atom. It has 6 protons, a mass number of 14 and must therefore have 8 (14 − 6) neutrons. $^{14}_{6}C$ is sometimes written as carbon-14.

Isotopes

D–C

- Isotopes are elements that have the same atomic number but different mass numbers.

B–A*

- Isotopes of an element have different numbers of neutrons in their atoms.

isotope	electrons	protons	neutrons
$^{1}_{1}H$	1	1	0
$^{2}_{1}H$	1	1	1
$^{3}_{1}H$	1	1	2

Electronic structure

D–C

- The same number of electrons occupies the space around the protons of the nucleus.
- Electrons occupy **shells**. The electron shell nearest to the nucleus takes up to 2 electrons. The second shell takes up to 8 electrons.

B–A*

- Each element has an **electron pattern** (**electronic structure**).
  – The **third shell** takes up to 8 electrons before the fourth shell starts to fill.
  – The **fourth shell** can take up to 18 electrons.
- The electronic structure of each of the first twenty elements can be worked out using:
  – the atomic number of the element
  – the maximum number of electrons in each shell.

a

the first shell takes up to 2 electrons

$^{4}_{2}He$

The distribution of electrons in **a** a helium atom **b** a carbon atom.

b

the second shell takes up to 8 electrons

$^{12}_{6}C$

The atomic number of lithium, Li, is 3. So the first 2 electrons of the lithium atom fill the first shell. The third electron goes into the second shell. The electronic structure is 2,1.

Questions

1 What's the mass of one proton?

2 What's the isotope that has 17 protons and 20 neutrons?

3 How many electrons does an atom of sodium have?

4 What's the element with electronic structure 2,8,2?

Ionic bonding

Ionic bonding

- A **metal atom** has extra electrons in its outer shell and needs to **lose** them to be stable. The electrons transfer from the metal atom to a non-metal atom to form a stable pair.
  - If an atom **loses electrons**, a **positive ion** is formed.
  - If an atom loses 1 electron, a (positive) $^+$ion is formed, e.g. $Na - e^- \longrightarrow Na^+$.
  - If an atom loses 2 electrons, a (positive) 2^+ ion is formed, e.g. $Mg - 2e^- \longrightarrow Mg^{2+}$.

- A **non-metal** atom has 'spaces' in its outer shell and needs to **gain** electrons to be stable. The electrons transfer to the non-metal atom from the metal atom to make a stable pair.
  - If an atom **gains electrons**, a **negative ion** is formed.
  - If an atom gains 1 electron, a (negative) $^-$ion is formed, e.g. $F + e^- \longrightarrow F^-$.
  - If an atom gains 2 electrons, a (negative) 2^- ion is formed, e.g. $O + 2e^- \longrightarrow O^{2-}$.

- During **ionic bonding**, the metal atom becomes a positive ion and the non-metal atom becomes a negative ion. The positive ion and the negative ion then attract one another.

- Sodium chloride **solution** conducts electricity. **Molten** (melted) magnesium oxide and sodium chloride conduct electricity.

- The '**dot and cross**' model describes ionic bonding.

Sodium chloride is a solid lattice made up of many pairs of ions held together by **electrostatic** attraction.
  - Sodium forms a positive sodium ion.
  - Chlorine forms a negative chloride ion.
  - The outer electron of sodium is transferred to the outer shell of the chlorine atom.

  - The sodium ion and the chloride ion are held together by attraction of opposite charges.

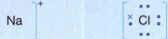

In **magnesium chloride**, magnesium needs to lose two electrons but chlorine can only gain one electron. So one magnesium atom needs two chlorine atoms to achieve an ionic bond and make magnesium chloride.

During the bonding in **magnesium oxide**, the magnesium atom loses two electrons to become a magnesium ion and the oxygen gains the two electrons to become an oxide ion.

  - The dot and cross model looks like this:

In **sodium oxide**, sodium only has one electron to lose, but oxygen needs to gain two electrons. Two sodium atoms are needed to bond with one oxygen atom.

- Each atom has either lost or gained the correct number of electrons to achieve a complete outer shell. It's called a **stable octet**.

- Molten liquids of ionic compounds can conduct electricity as the ions are free to move.

Questions

1 Explain how a negative ion is made from a neutral atom.

2 Explain the difference between how a metal atom and a non-metal atom transfers electrons.

3 Draw a 'dot and cross' model to show how an electron is transferred from a lithium atom to a fluorine atom.

4 Why do molten liquids of ionic compounds conduct electricity?

Covalent bonding

Covalent bonding

- Non-metals combine together by sharing **electrons**. This is called **covalent bonding**.

  A molecule of water is made up of three atoms: two hydrogen and one oxygen.
  – Oxygen has six electrons in its outer shell; it needs two more electrons to be complete.
  – Hydrogen atoms each have one electron in their only shell, so the oxygen outer shell is shared with each of the hydrogen electrons.
  – So each of the hydrogen atoms has a share of two more electrons making the shell full.

  A molecule of carbon dioxide is made up of three atoms: two oxygen and one carbon.
  – Carbon has four electrons in its outer shell; it needs four more electrons to be complete.
  – Oxygen atoms each have six electrons in their outer shell, so they each need two more electrons to be complete.
  – The oxygen outer shell is shared with two of the electrons of the carbon outer shell each.
  – So each of the oxygen atoms has a share of two more electrons making the shell full.

- Carbon dioxide and water don't conduct electricity because they are covalently bonded.

- The formation of simple molecules containing single and double covalent bonds can be represented by '**dot and cross' models**.

Predicting chemical properties

- Carbon dioxide and water are simple molecules with weak **intermolecular forces**.

- The chemical properties of carbon dioxide and water are related to their structure.
  – As they have weak intermolecular forces between the molecules, they're easy to separate, so the substances have low melting points.
  – As there are no free electrons available they don't conduct electricity.

weak intermolecular forces between the molecules

Group numbers

- This is how to tell which **group number** an element belongs to:
  – group 1 elements have 1 electron in the outer shell
  – group 7 elements have 7 electrons in the outer shell
  – group 8 elements have 8 electrons in the outer shell.

  Group 1   Group 7   Group 8

- This is how to tell which **period** an element belongs to:
  – electrons in only **one shell**, it's in the **first period**
  – electrons in **two shells**, it's in the **second period**
  – electrons in **three shells**, it's in the **third period**.

element	electrons	period
H	1	1
Li	2,1	2
Na	2, 8, 1	3

Questions

(Grades D-C)

1 Draw a model showing the bonding of a water molecule.

(Grades B-A*)

2 Draw a dot and cross model of the bonding in water.

(Grades D-C)

3 Sulphur has an electron pattern of 2, 8, 6. To which period does it belong?

(Grades B-A*)

4 There are only weak intermolecular forces between the molecules of carbon dioxide. Explain why this causes it to have a low melting point.

The group 1 elements

Properties of alkali metals

- When lithium, sodium and potassium react with water:
  - they float on the surface because their **density** is less than the density of water
  - hydrogen gas is given off
  - the metal reacts with water to form an **alkali** – the **hydroxide** of the metal
  - the reactivity of the alkali metals with water increases down group 1.

  Lithium reacts quickly and vigorously with water.
  lithium + water $\longrightarrow$ lithium hydroxide + hydrogen

  Sodium reacts very quickly and vigorously with water and forms sodium hydroxide.
  sodium + water $\longrightarrow$ sodium hydroxide + hydrogen

  Potassium reacts extremely vigorously with water and produces a lilac flame and forms potassium hydroxide.
  potassium + water $\longrightarrow$ potassium hydroxide + hydrogen

- Group 1 metals have **1 electron** in their **outer shell**, which is why they have similar properties.

- The word equation: sodium hydroxide + water $\longrightarrow$ sodium + hydrogen can be represented in symbols: $Na + H_2O \longrightarrow NaOH + H_2$ but the equation isn't balanced.
  - The balanced symbol equation is: $2Na + 2H_2O \longrightarrow 2NaOH + H_2$

$_3$Li	
$_{11}$Na	
$_{19}$K	
$_{37}$Rb	
$_{55}$Cs	
$_{87}$Fr	

- The reactivity with water increases down group 1.
  - We can predict the way that other alkali metals behave from the patterns so far.
  - We can see that rubidium reacts more vigorously with water than potassium.
  - We can predict the trends of the melting point, boiling point, appearance and electrical conductivity of rubidium and caesium from the trends of the other three.

Flame tests

- If you want to test the flame colours of the chemicals:
  - Put on safety goggles. Moisten a flame test wire with dilute hydrochloric acid.
  - Dip the flame test wire into the sample of solid chemical.
  - Hold the flame test wire in a blue Bunsen burner flame.
  - Record the colour of the flame in a table.

Explaining reactivity patterns

- Alkali metals have similar properties because when they react, their atoms need to lose one electron to form full outer shells. This is then a **stable electronic structure**.
  - When the atom loses one electron it forms an **ion**. The atom becomes charged.
  - It has more positive charges in its nucleus than negative electrons surrounding it, so the overall charge is positive. It has made a **positive ion**.
  - This can be represented by an equation. $Na - e^- \longrightarrow Na^+$

- Lithium loses its outer electron from its second shell. Sodium loses its outer electron from its third shell. The third shell is further away from the attractive 'pulling force' of the nucleus so the electron from sodium is more easily lost than the electron from lithium. Sodium is therefore more reactive than lithium.

- If electrons are lost, the process is called **oxidation**.

Questions

Grades D-C

1 Why does potassium float on water?
2 Explain why group 1 metals have similar properties.

Grades B-A*

3 What alkali will be made when rubidium reacts with water?
4 Why is sodium less reactive than potassium?

The group 7 elements

Group 7 trends

D–C

- The **physical appearance** of the halogens at room temperature is:
  – chlorine is a green gas
  – bromine is an orange liquid
  – iodine is a grey solid.

- Group 7 elements have similar properties:
  – they all have **7 electrons** in their **outer shell**.

fluorine 2, 7   chlorine 2, 8, 7

B–A*

  – melting points and boiling points increase down the group
  – when they react each atom gains one electron to form a
  negative ion with a stable electronic structure.

bromine (outer shell   iodine (outer shell
only shown) 7   only shown) 7

> Chlorine has an electronic structure of 2, 8, 7. It gains one electron to become 2, 8, 8.
> $Cl_2 + 2e^- \longrightarrow 2Cl^-$

- The nearer the outer shell is to the nucleus, the easier it is for an atom to gain one electron. The easier it is to gain the electron, the more reactive the halogen.

- If electrons are gained, the process is called **reduction**. In the ionic equation above, a molecule of chlorine gains two electrons to become two chlorine ions. The chlorine ion is a negative ion.

Halogens and reactivity

D–C

- Halogens react vigorously with alkali metals to make **metal halides**.

> When lithium reacts with chlorine, the metal halide made is lithium chloride.
> lithium + chlorine ⟶ lithium chloride

> When potassium reacts with iodine, the metal halide made is potassium iodide.
> potassium + iodine ⟶ potassium iodide

B–A*

- To construct a **balanced symbol equation**:
  – write down the symbols for the alkali metal and the halogen (the **reactants**)

  potassium K (alkali metal) chlorine, Cl_2 (halogen)

  – write down the formula for the **product**

  $K + Cl_2 \longrightarrow KCl$

  – balance number of molecules

  $2K + Cl_2 \longrightarrow 2KCl$

Displacement reactions of halogens

D–C

- The **reactivity** of the halogens decreases down the group.

- If halogens are bubbled through **solutions of metal halides**, there are two possibilities:
  – **no reaction**: if the halogen is less reactive than the halide in solution
  – a **displacement reaction**: if the halogen is more reactive than the halide in solution.

> Chlorine displaces the bromide to form bromine solution.
> chlorine + potassium bromide ⟶ potassium chloride + bromine (orange solution)

> Chlorine also displaces iodides from sodium iodide solution.
> chlorine + sodium iodide ⟶ sodium chloride + iodine (red-brown solution)

Questions

(Grades B–A*)

1 What's reduction?

(Grades D–C)

2 Write down the word equation for the reaction between potassium and bromine.

(Grades B–A*)

3 Write a balanced symbol equation for the formation of lithium bromide from its elements.

(Grades D–C)

4 Why doesn't iodine displace bromine from potassium bromide?

Electrolysis

The electrolysis of dilute sulphuric acid

- The key features of the **electrolysis** of dilute sulphuric acid are:
  - the **electrolyte** is a dilute solution of sulphuric acid
  - two **electrodes** are connected to a dc source of electric current, between 6 V and 12 V, and placed into the electrolyte
  - the electrode connected to the **negative terminal** is the **cathode**
  - the electrode connected to the **positive terminal** is the **anode**.

- When the current is switched on, bubbles of gas appear at both electrodes. Water splits into two ions: H^+ is the positive ion and OH^- is the negative ion.
  - H^+ is attracted to the negative cathode and discharged as hydrogen gas, H_2.
  - OH^- is attracted to the positive anode and discharged as oxygen gas, O_2.

- Twice the volume of hydrogen gas is given off as oxygen gas because the formula of the compound breaking up is H_2O.

Electrolysis of sulphuric acid in the laboratory.

Electrolysis of sodium chloride

- The reactions at the electrodes in the **electrolysis** of a dilute solution of **sodium chloride** are:
  - **at the cathode**: $2H^+ + 2e^- \longrightarrow H_2$
  - **at the anode**: $4OH^- - 4e^- \longrightarrow 2H_2O + O_2$

Electrolytic decomposition

- The key features in the production of aluminium by electrolytic decomposition are:
  - the use of molten aluminium oxide
  - aluminium is formed at the graphite cathode; oxygen is formed at the graphite anode
  - the anodes are gradually worn away by **oxidation**
  - the process requires a high electrical energy input.

- The word equation for the decomposition of aluminium oxide is:

  aluminium oxide $\longrightarrow$ aluminium + oxygen

carbon anodes

steel cathode

bauxite melt (molten AlO_3)

aluminium metal

- The electrode reactions in the **electrolytic** extraction of **aluminium** are:
  - **at the cathode**: electrons have been gained – an example of **reduction**
    $Al^{3+} + 3e^- \longrightarrow 3Al$
  - **at the anode**: electrons have been lost – an example of **oxidation**
    $2O^{2-} - 2e^- \longrightarrow 2O_2$

- The chemical **cryolite** is used to lower the melting point of aluminium oxide. Aluminium oxide requires large amounts of electricity to melt at very high temperatures, which is very expensive.

Questions

Grade D-C

1 The ratio of hydrogen gas to oxygen gas made during the electrolysis of water is 2:1. Explain why.

Grade B-A*

2 The H^+ ions are discharged at the cathode as H_2. Explain how.

Grade D-C

3 What happens to the anodes during the process of electrolysis of aluminium?

Grade B-A*

4 Why is cryolite added in the electrolysis of aluminium oxide?

Transition elements

Transition elements

- A **compound** that contains a transition element is often coloured:
  - copper compounds are blue
  - iron(II) compounds are pale green
  - iron(III) compounds are orange/brown.

- A transition and its compounds are often **catalysts**:
  - iron is used in the **Haber process** to make ammonia, which is used in fertilisers
  - nickel is used to harden the oils in the manufacture of margarine.

> **Top Tip!**
> A catalyst is an element or compound that changes the rate of a chemical reaction without taking part in the reaction. Catalysts are unchanged during the reaction.

Sodium hydroxide solution

- Sodium hydroxide solution is used to identify the presence of transition metal ions in solution:
  - Cu^{2+} ions form a blue solid
  - Fe^{2+} ions form a grey/green solid
  - Fe^{3+} ions form an orange **gelatinous** solid.

- These solids are metal hydroxide precipitates.

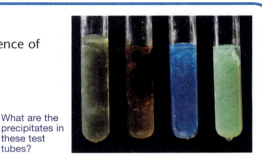

What are the precipitates in these test tubes?

Thermal decomposition

- If a transition metal carbonate is heated, it **decomposes** to form a metal oxide and carbon dioxide. On heating:
  - $FeCO_3$ decomposes forming iron oxide and carbon dioxide
  - $CuCO_3$ decomposes forming copper oxide and carbon dioxide
  - $MnCO_3$ decomposes forming manganese oxide and carbon dioxide
  - $ZnCO_3$ decomposes forming zinc oxide and carbon dioxide.

- The metal carbonates change colour during decomposition.

> **Top Tip!**
> The test for carbon dioxide is that it turns limewater milky.

- To write the balanced symbol equation for thermal decomposition:
  - write a word equation to establish the products of the reaction:

    copper carbonate $\longrightarrow$ copper oxide + carbon dioxide

  - assign symbols to the words:

    $$FeCO_3 \longrightarrow FeO + CO_2$$
    $$CuCO_3 \longrightarrow CuO + CO_2$$
    $$MnCO_3 \longrightarrow MnO + CO_2$$
    $$ZnCO_3 \longrightarrow ZnO + CO_2$$
    (these are balanced)

  - sometimes you need to balance the equation:

    $$Cu^{2+} + 2OH^- \longrightarrow Cu(OH)_2$$
    $$Fe^{2+} + 2OH^- \longrightarrow Fe(OH)_2$$
    $$Fe^{3+} + 2OH^- \longrightarrow Fe(OH)_3$$

Questions

Grades D–C

1 What colour are iron(III) compounds?
2 There is a difference between Fe^{2+} ions and Fe^{3+} ions. How would you show this using sodium hydroxide?

Grades B–A*

3 Write a word equation for the thermal decomposition of copper carbonate.
4 Write an ionic equation for the precipitation reaction between copper ions and hydroxyl ions.

Metal structure and properties

Properties of metals

- A property can be either **physical** or **chemical**.

- Physical properties of metals include:
  - having high thermal conductivity
  - being good conductors of heat
  - being malleable
  - being ductile
  - having high melting points and boiling points because of strong metallic bonds.

  Copper is often used for the base or the whole of saucepans because it has high thermal conductivity.

- Chemical properties of metals include:
  - resistance to attack by oxygen or acids.

  Copper is also resistant, which is another reason why it's used for saucepans.

  Aluminium has a low density and is used where this property is important, such as in the aircraft industry and also in modern cars.

- A **metallic bond** is a strong electrostatic force of attraction between close-packed positive metal ions and a 'sea' of **delocalised electrons**.

- Metals have high melting points and boiling points because a lot of energy is needed to overcome the strong attraction between the delocalised electrons and the positive metal ions.

Conductors and superconductors

- When metals conduct electricity, the electrons in the metal move. **Superconductors** are materials that conduct electricity with little or no resistance.

- When a substance goes from its normal state to a superconducting state, it no longer has a magnetic field. This is called the Meissner effect.

- The potential benefits of superconductors are:
  - loss-free power transmission
  - super-fast electronic circuits
  - powerful electromagnets.

The permanent magnet levitates above the superconductor.

- Metals conduct electricity because delocalised electrons within its structure can move easily.

- Superconductors only work at very low temperatures, so scientists need to develop superconductors that will work at 20 °C.

electrons from outer shells of metal atoms are free to migrate

The structure of a metal.

metal ions

Questions

1 Give two reasons why copper is good for making saucepans.

2 A metallic bond is a strong bond. Explain why.

3 What happens if a small permanent magnet is put above a superconductor?

4 What are the disadvantages of superconductors at present?

C3 Summary

Atoms and bonding

Atoms are neutral because they have the same number of electrons as protons. The electrons are arranged in a pattern or configuration. The outer shell of electrons needs to be full to be stable.

Atoms join together to make molecules or large crystal structures. There are two ways in which atoms can bond, by making ions or by sharing electrons.

Ions are made when atoms lose or gain electrons to make a full outer shell. Ions are either positive or negative.

Atoms can share electrons to make molecules containing two or more atoms. This bonding is called **covalent bonding**.

Periodic table

Group 1 metals react vigorously with water to make alkaline solutions, which are the hydroxide of the metal.

Group 7 elements are called the halogens. They have seven electrons in their outer shell.

The periodic table lists all elements in order of their **atomic number**. The elements of group 1 all have one electron in their outer shell.

The periodic table lists elements in groups. The elements have similar properties. The group that an element belongs to can be deduced from its electron pattern.

Electrolysis

Sulphuric acid decomposes to give hydrogen and oxygen. Hydrogen is made at the cathode and oxygen is made at the anode.

Electrolysis is the decomposition of a substance using electricity.

Aluminium is made by the electrolysis of bauxite. The mineral has to be purified before it's used. The aluminium is deposited at the cathode. Oxygen is given off at the anode. The graphite anodes are burned away in the process.

Transition metals and metal structure

Transition metals and their compounds are often catalysts. Their carbonates decompose on heating to give the **metal oxide** and **carbon dioxide**.

Metals **conduct electricity** easily because electrons move through the structure easily. At low temperatures some metals can become **super conductors**. These show little or no resistance when conducting electricity.

Transition metal compounds are usually coloured. The compounds often dissolve in water to make coloured solutions. The solutions react with sodium hydroxide to make coloured precipitates.

Speed

Measuring speed

Top Tip!
Remember! 1 km = 1000 m, 1 hour = 60 × 60 = 3600 seconds.

- The formula for speed is:

$$\text{average speed} = \frac{\text{distance}}{\text{time}} = \frac{d}{t}$$

We write 'average speed' because the speed of a car changes during a journey.

An aircraft travels 1800 km in 2 hours.

$$\text{Average speed} = \frac{1800}{2} = 900 \text{ km/h} = 900 \times \frac{1000}{3600} = 250 \text{ m/s}$$

- Increasing the speed means increasing the distance travelled in the same time.
  Increasing the speed reduces the time needed to cover the same distance.

- The formula for average speed can be rearranged to work out:
  – the distance travelled in a certain time   distance = average speed × time
  – how long it takes to travel a known distance $\text{time} = \dfrac{\text{distance travelled}}{\text{average speed}}$

D–C

B–A*

Distance-time graphs

- Distance-time graphs allow a collection of data to be shown. It is easier to interpret data when they are plotted on a graph than when they are listed in a results table.

- The **gradient** of a distance-time graph tells you about the speed of the object.
  A higher speed means a steeper gradient.

- In graph **a**, the distance travelled by the object each second is the same.
  The gradient is constant, so the speed is constant.
  In graph **b**, the distance travelled by the object each second increases as the time increases.
  The gradient increases, so there is an increase in the speed of the object.

- Speed is equal to the gradient of a distance-time graph; the higher the speed the steeper the gradient.

- A straight line indicates that the speed is constant. A curved line shows that the speed is changing.

The gradient of the graph is:

$$\text{gradient} = \frac{AC}{BC} = \frac{(20 - 10)}{(5 - 0)} = \frac{10}{5} = 2$$

This means speed = 2 m/s

D–C

B–A*

Questions

1 A car travels 600 m in 20 seconds. What is its average speed?

2 How long will it take to travel 560 km at an average speed of 80 km/h?

3 What can you say about the gradient of a distance-time graph for a car journey if **a** the car is travelling at a steady speed **b** the speed of the car is decreasing?

4 The gradient of a distance-time graph for a car is 25. What does this tell you about the motion of the car?

Changing speed

Speed-time graphs

D–C

- If the speed is increasing, the object is **accelerating**. If the speed is decreasing, the object is **decelerating**.

- The **area** under a speed-time graph is equal to the **distance travelled**.
  - The speed of car B in the graph is increasing more rapidly than the speed of car A, so car B is travelling further than car A in the same time.
  - The area under line B is greater than the area under line A for the same time.
  - The speed of car D is decreasing more rapidly than the speed of car C, so car D isn't travelling as far as car C in the same time.
  - The area under line D is smaller than the area under line C for the same time.

- The **gradient** (steepness) of a speed-time graph is the **acceleration**.

B–A*

- The formula for distance travelled is:

  distance travelled = area under speed-time graph

- If the speed is not changing uniformly the acceleration is not constant. The area under the graph is **estimated** to find the distance travelled.

> **Top Tip!**
> Don't confuse distance-time and speed-time graphs. Always look at the axes carefully.

Acceleration

D–C

- During a car journey, the speed of the car increases and decreases; it doesn't stay constant.

- The formula for measuring acceleration is: $\text{acceleration} = \dfrac{\text{change in speed}}{\text{time taken}}$

- A negative acceleration shows the car is slowing down, or decelerating.

A new car boasts a rapid acceleration of 0 to 108 km/h in 6 seconds.

A speed of 108 km/h is $\dfrac{108 \times 1000}{60 \times 60} = 30$ m/s

$\text{Acceleration} = \dfrac{\text{change in speed}}{\text{time taken}} = \dfrac{(30 - 0)}{6} = 5$ m/s^2

This means the speed of the car increases by 5 m/s every second.

> **Top Tip!**
> Make sure you know how to change km/h to m/s.

Circular motion

A moving object that changes direction moves tangentially to a circle or its arc.

B–A*

- A vehicle may go around a roundabout at a constant **speed** but it is **accelerating**! This is because its **direction** of travel is changing; it's not going in a straight line. The driver needs to apply a **force towards the centre** of the roundabout to change direction. This gives the vehicle an acceleration directed towards the centre of the roundabout.

- Any object moving along a circular path moves **tangentially** to the circle, or arc of a circle.

- **Velocity** is the speed of a moving object in a known direction.

- The formula $\text{acceleration} = \dfrac{\text{change in speed}}{\text{time taken}}$ can be rearranged to find:
  - a change in speed change in speed = acceleration × time
  - time taken $\text{time taken} = \dfrac{\text{change in speed}}{\text{acceleration}}$

> **Top Tip!**
> Make sure you can change the subject of the acceleration equation. Practise until you find it easy.

Questions

Grades D-C

1 What can be found from the area under a speed-time graph?

Grades B-A*

2 How can you find the acceleration of a car from its speed-time graph?

Grades D-C

3 Find the acceleration of a cat that goes from 0 to 5 m/s in 2 seconds when chasing a mouse.

Grades B-A*

4 A car accelerates from rest at 4 m/s^2 for 6 seconds. How fast is it then going?

Forces and motion

Force, mass and acceleration

- If the forces acting on an object are balanced, it's at rest or has a constant speed.
  If the forces acting on an object are unbalanced, it speeds up or slows down.

- The unit of force is the **Newton** (N).

- $F = ma$, where F = unbalanced force in N, m = mass in kg and a = acceleration in m/s^2.

  Marie pulls a sledge of mass 5 kg with an acceleration of 2 m/s^2 in the snow.
  The force needed to do this is: $F = ma$ $F = 5 \times 2 = 10$ N

- The equation $F = ma$ is used to find mass or acceleration if the resultant force is known.
  Professional golfers hit a golf ball with a force of approximately 9000 N.
  If the mass of the ball is 45 g, the acceleration during the very short
  time (about 0.005 milliseconds) of impact can be calculated.

  $$a = \frac{F}{m} = \frac{9000}{0.045} = 200\,000 \text{ m/s}^2$$

Top Tip!

Remember the correct units when using $F = ma$; F in N, m in kg, a in m/s^2.

- Forces occur in pairs. They:
  – are the same size
  – are in opposite directions
  – act on different objects.

When the two vans collide, the forces on each are of equal size but in opposite directions.

F_1 F_2

Top Tip!

Remember! These pairs of forces are *not* balanced forces.

Driving safely

- **Thinking time**, and therefore thinking distance, may increase if a driver is:
  – tired
  – under the influence of alcohol or other drugs
  – distracted or lacks concentration.

- The formula for **stopping distance** is:
  stopping distance = thinking distance + braking distance

- For safe driving, it is important to:
  – keep an appropriate distance from the car in front
  – have different speed limits for different types of road
  – slow down when road conditions are poor.

- **Braking distance** may increase if:
  – the road is icy or wet
  – the car has poor brakes or bald tyres
  – the car is travelling fast.

Top Tip!

For an alert driver, thinking time (or **reaction time**) is about 0.7 seconds.

- Factors affecting braking distance are:
  – The greater the **mass** of a vehicle the greater its braking distance.
  – The greater the **speed** of a vehicle the greater its braking distance.
  – When the brakes are applied the brake pads are pushed against the disc.
  This creates a large friction force that slows the car down. Worn brakes
  reduce the friction force, increasing the braking distance.
  – Worn **tyres** with very little **tread** reduce the grip of the wheels on a
  slippery road, leading to skidding and an increase in braking distance.

Questions

(Grades D-C)

1 Calculate the unbalanced force needed to give a car of mass 1200 kg an acceleration of 5 m/s^2.

(Grades B-A*)

2 Show how the gravitational forces acting on the Earth and the Moon fit the rules for force pairs.

(Grades D-C)

3 Tim's reaction time is 0.7 seconds. Calculate his thinking distance when travelling at 20 m/s.

(Grades B-A*)

4 Explain why worn brakes can increase braking distance.

Work and power

Work

D–C

- **Work** is done when a force moves in the direction in which the force acts.

- The formula for work done is:
  work done = force × distance moved (in the direction of the force)

  If a person weighs 700 N, the work he does against gravity when he jumps 80 cm is:
  work done = force × distance moved = 700 × 0.8 = 560 J

B–A*

- The formula for work done can be rearranged to work out:

  – force $force = \dfrac{work\ done}{distance}$

  – distance $distance = \dfrac{work\ done}{force}$

- When car brakes are applied the brake pads are pushed against a disc. This creates a large **friction** force that slows the car down.

- A car loses all its **kinetic energy** when it stops. The kinetic energy is transferred mainly into heat by the brakes.

- kinetic energy lost = work done by brakes

- braking distance = $\dfrac{work\ done\ by\ brakes}{braking\ force}$

Calculating power

D–C

- The formula for **power** is: $power = \dfrac{work\ done}{time\ taken}$

- A person's power is greater when they run than when they walk.

Top Tip!
Always check the units when doing calculations on work, power and energy.

B–A*

- The equation for power can be rearranged to work out:
  – work done   work done = power × time
  – time taken $time\ taken = \dfrac{work\ done}{power}$

  When the Eurostar travels at maximum speed, its power is 2 MW. The amount of work done, or energy transferred, in 2 hours is calculated by:
  work done = power × time = 2 000 000 × (120 × 60) = 14 400 000 000 J (14 400 MJ)
  At maximum power, 14 400 MJ of energy would be transferred to other forms during a 2 hour journey. The Eurostar operates at maximum power for only a short part of the journey.

Fuel

D–C

- Fuel is expensive and a car with high fuel consumption is expensive to run.

- Fuel pollutes the environment.
  – Car exhaust gases, especially carbon dioxide, are harmful.
  – Carbon dioxide is a major source of **greenhouse gases**, which contribute to climate change.

Questions

Grades D-C

1 How much work does Lee do when he pushes a car 20 m with a force of 80 N?

Grades B-A*

2 Find the braking force if a car loses 250 000 J of KE when braking in a distance of 50 m. What happens to the KE lost?

Grades D-C

3 Find the power of a crane that can lift a load of 500 N through a height of 12 m in 15 seconds.

Grades B-A*

4 How much work is done by a 3 kW motor when it lifts a heavy load in 1 minute? Suggest why in practice the motor will actually need to do more work than you have calculated.

Energy on the move

Fuel consumption

- **Fuel consumption** data are based on ideal road conditions for a car driven at a steady speed in urban and non-urban conditions.

car	fuel	engine in litres	miles per gallon (mpg)	
			urban	non-urban
Renault Megane	petrol	2.0	25	32
Land-Rover	petrol	4.2	14	24

- Factors that affect the fuel consumption of a car are:
  - the amount of energy required to increase its kinetic energy
  - the amount of energy required for it to do work against friction
  - its speed
  - how it is driven, such as excessive acceleration and deceleration, constant braking and speed changes
  - road conditions, such as a rough surface.

Top Tip!

Make sure you can interpret fuel consumption tables.

Kinetic energy

- **Kinetic energy** increases with:
  - increasing **mass**   – increasing **speed**.

- The **braking distance** of a car increases with increasing speed, but not proportionally.

- The formula for kinetic energy is: kinetic energy = $\frac{1}{2} mv^2$
  where m = mass (in kg), v = speed (in m/s).

  If a car has a mass of 1000 kg, its kinetic energy:
  - at 20 m/s is $\frac{1}{2} mv^2 = \frac{1}{2} \times 1000 \times (20)^2 = 200\ 000$ J
  - at 40 m/s is $\frac{1}{2} mv^2 = \frac{1}{2} \times 1000 \times (40)^2 = 800\ 000$ J

- When a car stops its kinetic energy changes into heat in the brakes, tyres and road. This can be shown by the formula: work done by brakes = loss in kinetic energy

- The change in kinetic energy can be shown in the formula:
  braking force × braking distance = loss in kinetic energy

- When the **speed** of the car **doubles**, the **kinetic energy** and the **braking distance quadruple**.

Electrically powered cars

- Exhaust fumes from petrol-fuelled and diesel-fuelled cars cause serious pollution in towns and cities.

- Battery-driven cars do not pollute the local environment, but their batteries need to be recharged. Recharging uses electricity from a power station. Power stations pollute the local atmosphere and cause acid rain.

This hybrid electric car has solar panels on its roof that convert sunlight into additional power to supplement its battery.

Questions

Grades B–A*

1 Suggest why a car uses more fuel per kilometre when carrying a heavy load.

2 Suggest three things a driver can do to improve the fuel consumption of his car.

3 **a** Calculate the kinetic energy of a car of mass 1200 kg going at 10 m/s. **b** Calculate its braking distance if the braking force is 5000 N.

Grades D-C

4 Electric milk floats have been used for many years. Suggest why more people don't use electric cars.

Crumple zones

Car safety

D–C

- On impact:
  - **crumple zones** at the front and rear of the car absorb some of its energy by changing shape or 'crumpling'
  - a **seat belt** stretches a little so that some of the person's kinetic energy is converted to elastic energy
  - an **air bag** absorbs some of the person's kinetic energy by squashing up around them.

An air bag and seat belt in action during a car accident. What parts of the man are protected?

- All these safety features:
  - absorb energy
  - change shape
  - reduce injuries.

- Active safety features include:
  - **ABS brakes** (anti-lock brake system) which give stability and maintain steering during hard braking. The driver gets the maximum braking force without skidding and can still steer the car.
  - **Traction control** which stops the wheels on a vehicle from spinning during rapid acceleration.
  - A car **safety cage** which is a **rigid** frame that prevents the car from collapsing and crushing the occupants in a roll-over crash.
  - Crumple zones at front and rear which keep damage away from the internal safety cage.

- Passive safety features help the driver to concentrate on the road. These include:
  - cruise control
  - electric windows
  - paddle shift controls
  - adjustable seating.

Top Tip!

ABS brakes don't stop a car more quickly. They give improved control and prevent skidding.

B–A*

- In the **ABS** system, wheel-speed sensors send information to a computer about the rotational speed of the wheels. The computer controls the pressure to the brakes, via a pump, to prevent the wheels locking up. This increases the braking force (F) just before the wheels start to skid.
  - kinetic energy lost = work done by the brakes
  $$\tfrac{1}{2}mv^2 = F \times d$$
  where m = mass of car v = speed of car before braking d = braking distance.
  - If F increases, d decreases and the braking distance decreases.

- **Active safety features** have a more immediate effect in saving lives in an accident.

- **Passive safety features** contribute to safe driving but do not directly affect the safety of a car driver and passengers in an accident.

Reducing injury

B–A*

- To minimise injury, **forces** acting on the people in a car during an accident must be minimised.

- force = mass × acceleration
  Force can be reduced by reducing the **acceleration** by:
  - increasing stopping or collision time
  - increasing stopping or collision distance.

- Safety features that do this include:
  - crumple zones
  - seat belts
  - air bags
  - crash barriers
  - escape lanes.

Questions

Grades D-C

1 Suggest two forms of energy that a car's kinetic energy may be changed to in an accident.

2 Explain why cruise control is a passive safety feature.

Grades B-A*

3 ABS brakes apply a force of 8000 N to stop a car of mass 1000 kg moving at 20 m/s. Find the braking distance.

4 Explain how increasing collision time reduces the injury risk in a crash.

Falling safely

Falling objects

- All objects fall with the same acceleration as long as the effect of **air resistance** is very small.

- The size of the air resistance force on a falling object depends on:
  - its cross-sectional area – the larger the area the greater the air resistance
  - its speed – the faster it falls the greater the air resistance.

- Air resistance has a significant effect on motion only when it is large compared to the weight of the falling object.

Balls A and B fall with the same speed.

air resistance = 0.5 N   air resistance = 0.5 N

acceleration = 9.5 m/s²   A   B   acceleration = 5 m/s²

weight = 10 N   weight = 1 N

- The speed of a **free-fall** parachutist changes as he falls to Earth.
  - In picture 1, the weight of the parachutist is greater than air resistance. He accelerates.
  - In picture 2, the weight of the parachutist and air resistance are equal.
    The parachutist has reached **terminal speed** because the forces acting on him are **balanced**.
  - In picture 3, the air resistance is larger than the weight of the parachutist. The parachutist slows down and air resistance decreases.
  - In picture 4, the air resistance and weight of the parachutist are the same. He reaches a new, slower terminal speed.

1 600 N

2 600 N / 600 N

3 1000 N / 600 N

4 600 N / 600 N

- True free-falling objects fall under the influence of gravity only. They:
  - don't experience air resistance
  - accelerate downwards at the same rate irrespective of mass or shape.

Top Tip!

The net force acting on a parachutist = weight – air resistance.

Terminal speed

- Look at the parachutist in the panel above.
  - In picture 1, the parachutist **accelerates**, displacing more air **molecules** every second. The **air resistance** force **increases**. This **reduces** his **acceleration**. So, the higher the speed, the more air resistance.
  - In picture 2, the parachutist's **weight** is **equal to** the **air resistance**; the forces on him are balanced. He travels at a constant speed – **terminal speed**.
  - In picture 3, when the **parachute** opens, the **upward** force **increases** suddenly as there is a much larger surface area, displacing more air molecules every second. So, the larger the area, the more air resistance. The parachutist **decelerates**, displacing fewer air molecules each second, so the **air resistance** force **decreases**.
  - In picture 4, the parachutist reaches a new **slower terminal speed** when his **weight** is **equal to** the **air resistance** once more, so he lands safely.

- Drag racers and the Space Shuttle use parachutes to slow them down rapidly.

Questions

1 Name the two forces acting on a parachutist. What can you say about these forces **a** at the start of the descent **b** when terminal speed is reached **c** when the parachute opens?

2 Explain, in terms of forces, why a free-fall parachutist doesn't keep accelerating at 10 m/s² before opening his parachute.

3 Explain why air resistance increases with the speed of a falling object.

The energy of theme rides

Energy transfers

D–C

- A bouncing ball converts gravitational potential energy to kinetic energy and back to gravitational potential energy. It does not return to its original height because energy is transferred to other forms such as **thermal energy** and **sound energy**.

- In the diagram, the gravitational potential energy at D is less than at A.

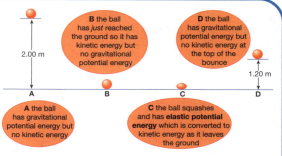

A the ball has gravitational potential energy but no kinetic energy

B the ball has *just* reached the ground so it has kinetic energy but no gravitational potential energy

C the ball squashes and has **elastic potential energy** which is converted to kinetic energy as it leaves the ground

D the ball has gravitational potential energy but no kinetic energy at the top of the bounce

The stages of energy transfer when a ball is dropped.

Calculating gravitational potential energy

B–A*

- The formula for gravitational potential potential energy is:
  gravitational potential energy (PE) = mgh
  where m = mass in kg h = vertical height moved in m g = gravitational field strength in N/kg
  (on Earth, g = 10 N/kg)

  If a ball of mass 50 g is dropped from a height **A** of 2 m and bounces up to a height **B** of 1.2 m, the energy wasted in the bounce is:
  PE lost between **A** and **B** = mgh = 0.05 × 10 × (2.00 − 1.20) = 0.4 J

- When a skydiver reaches **terminal speed,** their kinetic energy ($\frac{1}{2}mv^2$) has a maximum value and remains constant. The PE lost as they fall is used to do **work** against friction (**air resistance**).
  – When terminal speed is reached it can be shown as: change in PE = work done against friction

How a roller coaster works

D–C

- A roller coaster uses a motor to haul a train up in the air. The riders at the top of a roller coaster ride have a lot of gravitational potential energy.

- When the train is released, it converts **gravitational potential energy** to **kinetic energy** as it falls. This is shown by the formula:
  loss of gravitational potential energy (PE) = gain in kinetic energy (KE)

> **Top Tip!**
> Remember! Energy is always conserved.

- Each peak is lower than the one before because some energy is transferred to heat and sound due to friction and air resistance. This is shown by the formula:
  PE at top = KE at bottom + energy transferred (to heat and sound) due to friction

- kinetic energy = $\frac{1}{2}mv^2$
  – If **speed doubles, KE quadruples** (KE $\propto v^2$).   – If **mass doubles, KE doubles** (KE $\propto m$).

Mass and weight

B–A*

- The force of attraction on a mass due to gravity is called **weight**.
  – Weight = mg where m = mass of object g = gravitational field strength.
  Sadaf has a mass of 54 kg. She weighs 54 × 10 = 540 N on Earth.

> **Top Tip!**
> Don't confuse mass (in kg) and weight (in N).

Questions

[Take g = 10 N/kg.]

(Grades D-C)
1 Describe the energy changes for a girl on a swing.

(Grades B-A*)
2 Calculate the gravitational potential energy gained by a 50 kg mass when raised by a height of 8 m.

(Grades D-C)
3 Why does a roller coaster travel more slowly at the top than at the bottom of each hill?.

(Grades B-A*)
4 Tom weighs 600 N on Earth and 100 N on the Moon.
  a What is his mass? **b** What is the gravitational field strength on the Moon?

P3 Summary

Distance-time graphs

Speed is equal to the gradient of a distance-time graph.

Speed and acceleration

$$\text{speed} = \frac{\text{distance}}{\text{time}}$$

$$\text{acceleration} = \frac{\text{change in speed}}{\text{time}}$$

Speed-time graphs

The **gradient** of a speed-time graph is the **acceleration**.

distance travelled = area under a speed-time graph

Forces and motion

Force, mass and acceleration are linked by the equation:
$F = ma$

Forces always occur in **pairs** that:
– are the same size
– act in opposite directions
– act on different objects.

Forces can make things speed up or slow down.

When a car stops the:
total stopping distance = thinking distance + braking distance
An accident is more likely if the braking distance increases. This can be due to increased speed or worn tyres.

Work, energy and power

Work is done when a force moves an object.
$W = F \times d$
Work is measured in joules (J).

Roller coasters use **gravitational potential energy** (PE) as the source of movement.
When an object falls, it converts PE to KE.

Power is a measure of how quickly work is done.
$P = \frac{W}{t}$
Power is measured in watts (W).

kinetic energy (KE) $= \frac{1}{2}mv^2$
When a car brakes: work done
brakes = KE lost $f \times d = \frac{1}{2}mv^2$

weight = mass × gravitational field strength

Modern cars have lots of safety features that absorb energy when the cars stop:
– crumple zones
– seatbelts
– air bags.
These reduce forces by:
– increasing collision time
– increasing collision distance
– decreasing acceleration.

Falling safely

Objects reach a **terminal speed** because:
– higher speed = more drag
– larger area = more drag
– weight (or driving force)
  = drag, so forces are balanced (speed constant).

Falling objects get faster as they fall. They're pulled towards the centre of the Earth by their weight (**gravity**).

A parachute provides a large upward force, bigger than the weight of the parachutist. This slows him down. As he slows down the air resistance gets less until it equals his weight. He then falls at a terminal speed and lands safely.

Who planted that there?

The structure of a leaf

D–C

- When you cut through a leaf and look at it under a microscope you'll see many different cells.

- A **palisade** cell contains many **chloroplasts**. Photosynthesis occurs at a high rate in these cells and lots of sugars and starch are produced.

A stained section through a green leaf, magnified.

cuticle (wax layer)

upper epidermis

palisade cell

palisade layer

vein

green chloroplasts

spongy mesophyll cell

spongy mesophyll layer

air space

lower epidermis

guard cell   stoma (pore)   guard cell

Gas exchange in a leaf

D–C

- **Carbon dioxide** enters and **oxygen** exits a leaf through the **stomata** (one pore is called a stoma). The gases move by **diffusion**.

Photosynthesis

B–A*

- The structure and arrangement of **cells** are adapted for maximum efficiency of **photosynthesis**.
  - The **epidermis** is thin and transparent, allowing light through to inner cells.
  - The **palisade** cells contain large numbers of chloroplasts to absorb lots of light energy.
  - The **chloroplasts** are arranged mainly down the sides of the palisade cells, allowing some light to reach the **mesophyll cells**.
  - **Air spaces** between mesophyll cells allow gases to diffuse easily and reach all cells.
  - The mesophyll cells are small and irregular. This increases their surface area to volume ratio so large amounts of gases can enter and exit.

D–C

- A **leaf** is adapted for photosynthesis by:
  - being **broad** so it has a large **surface area** to absorb light
  - being **thin** so gases don't have far to travel and light can reach all the way through it
  - having **chlorophyll** in most of its cells
  - having a **network** of specialised **cells in veins** to support it and carry water and sugars to different parts of the plant
  - having **stomata** to allow **gas exchange**: carbon dioxide to **diffuse** into it and oxygen to diffuse out of it.

Questions

Grades D-C

1 Which type of leaf cell contains the most chloroplasts?

Grades B-A*

2 Explain why there are air spaces between mesophyll cells.

Grades D-C

3 Explain why a plant leaf is broad.

4 What is the purpose of stomata?

Water, water everywhere

Osmosis

- **Osmosis** is the **diffusion of water** across a **partially permeable membrane** from an area of high water concentration to an area of low water concentration.

- Water will move into cells by osmosis when they are placed in water. Cells will lose water by osmosis if they are placed in strong sugar solution.

Top Tip!

Remember osmosis is about **water movement** not about salt or sugar molecules moving.

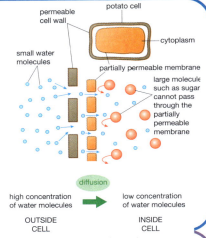

Osmosis in a plant cell.

- Potato chips placed in salt water become **flaccid** (soft and floppy). This is because there's a higher concentration of water molecules inside the potato cells than outside.

- When water **leaves** a plant cell by osmosis, the cell contents **shrink** and there's **less** water **pressure** against the cell wall. The cell contents are **plasmolysed** and the cell collapses.

- When water **enters** a plant cell it **swells up** and there's an **increase** in water pressure against the cell wall. This **turgor pressure** causes the cell to become **turgid** (hard and rigid).

- A **leaf** is adapted for **photosynthesis** so water easily evaporates and escapes from it. Plants need to reduce their water loss by having **few stomata**, which are mainly in the **lower epidermis**. Plants can also open and close their stomata.

- If a blood cell of an **animal** is placed in salt water or pure water it loses or gains water by osmosis. Since it has no cell wall for support it shrinks (becomes **crenate**) or bursts (**lysis**).

How a stomata opens.

Transpiration

- **Root hairs** are long and thin and have a large **surface area** to absorb lots of water from the soil by osmosis.

- The evaporation of water from leaves is called **transpiration**. It's useful because:
  - evaporation of water **cools** a plant
  - it brings **water** to the leaves for **photosynthesis**
  - a cell full of water gives **support**
  - the water moving up the stem carries useful **dissolved minerals**.

- To prevent too much water evaporating from the leaf it has a **waxy** covering called the **cuticle**, and **stomata** mainly on its shaded underside.

- The inelastic cell walls of a plant also help to support it.

Questions

Grades D-C

1 What name is given to the movement of water in and out of cells?

Grades B-A*

2 Describe the change you would see using a microscope if onion cells were placed in a strong salt solution.

3 Explain how osmosis is used to close the guard cell.

Grades D-C

4 Explain why stomata are mainly found on the underside of the leaf.

Transport in plants

Transport in plants

D–C

- **Xylem** and **phloem** are specialised cells that form the **transport system** in a plant. They form continuous **vascular bundles** from the roots to the stems and leaves.

- **Xylem cells** carry **water** and **minerals from the roots** to the leaves for photosynthesis. Some water evaporates and escapes by **transpiration** from the leaves.

- **Phloem cells** carry dissolved **food** such as **sugars from the leaves** to other parts of the plant. This movement is called **translocation**. The sugars can be used for growth or stored as starch.

Top Tip!

It's easy to get confused between xylem and phloem cells. Remember, phloem carries food.

Horizontal cross-sections through a broad bean plant to show the arrangement of xylem and phloem.

B–A*

- **Xylem cells** have thick, strong cell walls that contain a chemical called **lignin**.

- The **xylem vessels** are made of dead xylem cells joined to form a long hollow tube. The hole in the middle is called the **lumen**.

- **Phloem cells** also make long thin columns but the cells are still alive.

Increasing transpiration

D–C

- A **high rate** of **transpiration** happens when:
  - light intensity increases
  - temperature increases
  - air movement (wind) increases
  - humidity (amount of water in the atmosphere) falls.

B–A*

- When **light intensity** increases, the stomata open allowing more water to escape.

- As the **temperature** increases, the random movement of water molecules increases and more water escapes.

- **Wind** causes more water molecules near stomata to be removed. This increases evaporation and diffusion of water from inside the leaf.

- In **dry conditions** there's a very low concentration of water molecules outside the leaf. This causes more diffusion of water from inside the leaf to the outside.

Questions

Grades D-C

1 Name the vessels that carry water and minerals up the stem.

Grades B-A*

2 Explain why xylem vessels are strong enough to support a tree.

Grades D-C

3 Name two things that increase the rate of transpiration.

Grades B-A*

4 Explain how increased light intensity increases transpiration rate.

Plants need minerals too

Minerals

- Each mineral is used by a plant for different things.
  - **Nitrates** are needed to make **proteins** for growth.
  - **Phosphates** are used in **respiration** (releasing energy) and are needed for growth, especially in roots.
  - **Potassium** compounds are used in **respiration** and in **photosynthesis**.
  - **Magnesium** compounds are also needed in **photosynthesis**.

- A **mineral deficiency** in a plant is easy to detect and correct. A mineral is needed only in a very small amount.

no nitrates ➝ • poor growth
• yellow leaves

no phosphate ➝ • poor root growth
• discoloured leaves

no potassium ➝ • poor fruits and flowers
• discoloured leaves

no magnesium ➝ • yellow leaves

Symptoms of different mineral deficiencies in a plant.

- This is what happens to the minerals the plant takes in through fertilisers.
  - **Nitrogen** in nitrates is used to make **amino acids**. Amino acids are joined together to make different proteins such as **enzymes**.
  - **Phosphorus** in phosphates is used to make **cell membranes** and **DNA**. DNA carries genetic information.
  - **Potassium** is used to help make some enzymes. Enzymes speed up chemical reactions such as **photosynthesis** and **respiration**.
  - **Magnesium** is used to make **chlorophyll** molecules.

Active transport

- An increase in the uptake of minerals by a plant is matched by an increase in its **respiration rate**. This shows that **energy** is necessary for the uptake of minerals. The minerals are absorbed against a **concentration gradient**.

- A plant absorbs different minerals in different amounts and **selects** the minerals it needs.

- Special **carrier** molecules take the **mineral ions** across plant cell membranes by a process called **active transport**. Different carriers take different minerals.

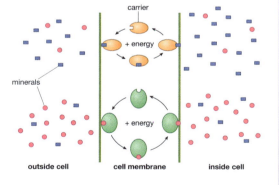

carrier

+ energy

+ energy

minerals

outside cell   cell membrane   inside cell

Active transport in a plant.

Top Tip!

Active transport needs energy from respiration but osmosis and diffusion don't.

Questions

Grades D-C

1 Name two minerals needed for photosynthesis.
2 Which mineral would you give a plant with poor root growth?

Grades B-A*

3 Which mineral is used to make chlorophyll?
4 Describe two differences between diffusion and active transport.

Energy flow

Food pyramids

- The links in food chains and food webs show different **trophic levels**. Energy is passed from one organism to another and each organism is at a different trophic level.

- The numbers of organisms at different trophic levels can be counted and the information shown in a **pyramid of numbers**.

- An animal that eats plants is called a **herbivore** and an animal that eats other animals is a **carnivore**.

- A **pyramid of biomass** is a better way of showing trophic levels because the mass of the organisms is used.

- As energy flows along a food chain some is used up in growth. At each trophic level about 90% of the energy is transferred into other less useful forms, such as heat from respiration and egestion of waste.

  a a woodland food chain doesn't show a pyramid shape because only one large tree supports all the other organisms **b** a pyramid of biomass of the same food chain.

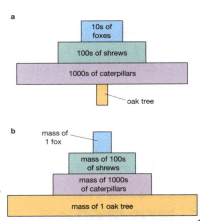

Biomass

- Because each trophic level 'loses' up to 90% of the available energy, the length of a food chain is limited to a small number of links.

- The shape of a **pyramid of biomass** shows that the energy decreases with increasing trophic levels.

- **Biomass** can be eaten; fed to other livestock (an inefficient transfer of energy); used as a source of seeds for next year (doesn't directly supply energy but saves costs); or used as a biofuel.

- To calculate the **efficiency of energy transfer** use this formula:

$$\text{efficiency} = \frac{\text{energy used for growth (output)}}{\text{energy supplied (input)}}$$

If there's 3056 kJ of energy in grass and only 125 kJ is used for a cow's growth:

$$\text{efficiency} = \frac{125}{3056} = 0.04 \text{ or } 4\%$$

This shows that the energy transfer to humans from beef is very inefficient.

Biofuels

- The energy from **biomass** can be used as **alternatives** to fossil fuels.
  - Fast-growing trees such as **willow** can be burnt in power stations. Because they are fast growing there's always a good supply.
  - Brazil produces a lot of **sugar cane** which is fermented using yeast to make alcohol. The process uses anaerobic respiration. The alcohol is mixed with petrol to make 'Gasohol', a fuel for cars.

- Here are some advantages of biofuels.
  - They're **renewable** by growing more plants or collecting methane.
  - They contribute less to air pollution.
  - Biofuels help towards a country's energy self-reliance.

Questions

Grades D-C

1 Describe the difference between a pyramid of numbers and a pyramid of biomass.

Grades B-A*

2 Calculate the efficiency if there's 2000 kJ of energy in grass and 40 kJ is used for growth of a cow.

Grades D-C

3 Explain how sugar cane can be used to fuel a car.

Grades B-A*

4 Explain why biofuels are renewable energy sources.

Farming

Intensive farming

- **Intensive farming** produces as much food as possible in the space available. Although it's an efficient method, it raises **ethical issues** such as cruelty to animals, and **environmental concerns** as the pesticides can pollute the land.

- Pesticides can build up to lethal doses in the food chain.
  - Pesticides from farmland get into lakes (0.02 ppm).
  - Microscopic life absorbs small amounts of the pesticide (5 ppm).

  - Fish eat large amounts of microscopic life and the pesticides build up inside them (200 ppm).
  - Grebes eat fish and the pesticide builds up killing them.

- Intensive farming is extremely **efficient** as more energy is **usefully transferred**. This is because:
  - there are fewer weeds and pests in crops
  - less heat is lost from animals kept in sheds and their movement is restricted.

Hydroponics

- A **hydroponics system** can be used to grow lettuce and tomato plants. The system doesn't use soil, so there's less chance of disease or pests affecting the plants. The plant roots are specially treated in water that contains the required amounts of fertiliser and oxygen. The system is also useful in countries with poor soil or little water to irrigate fields.

- A hydroponics system has many **advantages**.
  - Mineral supply is controlled and unused minerals are recycled, reducing cost.
  - It's set up under cover so there's better control of external conditions and disease.

- But there are some **disadvantages** too.
  - Manufactured fertilisers must be bought and tall plants such as tomatoes need support.

Some methods of organic farming

- **Weeds** are removed to reduce competition for light and minerals.

- **Seeds** are sown at **different times** so crops are ready at different times.

- **Beans** are grown to put nitrogen back into the soil. They're called **nitrogen-fixing plants**.

- **Manure** and **compost** are used in place of artificial fertilisers.

- **Crop rotation** is used so the same crops don't grow in the same place each year.

- Other **animals** can be used to kill pests, such as ladybirds that eat aphids. This saves on the use of harmful chemicals, but is often very slow to work and takes animals out of a food chain.

- Organic farming is a good idea in countries that *aren't* short of food. Other countries that have a poor climate can't produce enough organic food so they do need artificial fertilisers.

- The type of farming practised is a balance between **cost** and **suitability**.
  - In many developed countries it's fashionable to want organic food since it's seen to be healthier and people can afford to pay more.
  - In less well developed countries the farmers try to grow enough food, by any means possible, to survive.

Questions

Grades D-C

1 Suggest one way in which intensive farming is cruel to animals.

Grades B-A*

2 Suggest one advantage of hydroponics.

Grades D-C

3 Write down one way organic farmers can replace nitrogen lost from soil.

Grades B-A*

4 Suggest one disadvantage of organic farming.

Decay

Decay

- The remains of dead and decaying plants and animals are called **detritus**.

- Animals such as earthworms, maggots and woodlice depend on detritus
  for their food and are called **detritivores**.
  – Detritivores break down detritus into small pieces,
    which increases the surface area and so speeds up
    decay.
  – Detritivores are important as they **recycle**
    chemicals from dead plants and animals.

- Another method of using decay is for **compost**.
  These are the ideal conditions for speeding up the
  composting process:
  – **warmth**, such as placing a compost bin in the Sun
  – **moisture**
  – good **aeration**, such as regular mixing of the contents to allow oxygen in.

Two food chains to show how dead animals and decaying plant material are recycled.

dead animal → blowflies and blowfly maggots → common frog → grass snake

decaying leaves → earthworms → blackbird → sparrowhawk

- As **microorganisms** living in a compost bin **respire**, heat is transferred which warms up the
  compost, which in turn speeds up decay.

- Decay depends on **enzyme action**, so conditions that speed up enzyme action also speed up
  decay. The **optimum temperature** (the temperature at which enzymes work best) is
  approximately 37 °C for bacteria and 25 °C for fungi.

- **Aerobic** bacteria break down dead detritus quickly. They need a good supply of oxygen to carry
  out aerobic respiration.

- In a lack of oxygen **anaerobic** bacteria grow. These produce acid conditions that slow down decay.

- A fungus that feeds off dead and decaying material is called a **saprophyte**. Its digestive enzymes
  are released on to the food and break it down into simple soluble substances so the digested food
  is absorbed.

Preserving foods

- **Canning** involves heating food to kill bacteria, then sealing the food in a can
  to stop bacteria getting in.

- **Cooking** kills bacteria. Food poisoning can be caused by food being undercooked.

- **Vinegar** is an acid. Very few bacteria can grow in acid conditions.
  Pickled eggs and chutney are preserved in this way.

- **Drying** foods, such as cereals, works because bacteria
  and fungi can't grow without water.

- **Freezing** kills bacteria or slows their growth;
  cooling just slows down the growth.

- **Adding sugar** or **salt** kills some bacteria and fungi
  and stops the growth of others.

Top Tip!

Remember any preserving method stops microbes growing because it takes away warmth, moisture or oxygen.

Questions

Grades D-C

1 Name three detritivores.

Grades B-A*

2 What is meant by the term 'optimum temperature'?

3 Explain what's meant by the term 'saprophyte'.

Grades D-C

4 Explain why drying prevents food decay.

Recycling

The carbon cycle

- **Carbon dioxide** is a compound that contains carbon.

- Carbon dioxide is **taken out** of the atmosphere by plants to use in **photosynthesis**. The carbon is then passed along the food chain when animals feed.

- Carbon dioxide is **put back** into the atmosphere by:
  – plants and animals **respiring**
  – bacteria and fungi in soil respiring when they **decompose** organisms

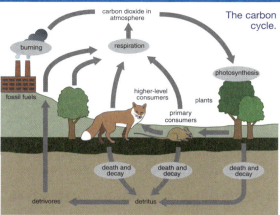

The carbon cycle.

  – **burning** fossil fuels such as coal and oil
  – **erupting** volcanoes and forest fires.

- Carbon is also **recycled at sea**. The shells of marine organisms such as molluscs contain carbonates. When the organisms die their shells turn into a **sedimentary** rock called **limestone**. Limestone is attacked by **acid rain**, which weathers the rock, and carbon dioxide is released.

The nitrogen cycle

- **Nitrogen** is also recycled naturally. The atmosphere is 78% nitrogen. Plants and animals need nitrogen for growth, but can't use it directly because it is **unreactive**.

- **Decomposers** break down proteins in dead bodies and **urea** into **ammonia**. **Nitrifying bacteria** convert ammonia into **nitrates**.

- **Nitrogen-fixing bacteria** are found in soil or root nodules of plants such as beans. They 'fix' nitrogen by converting it into ammonia or nitrates and then into amino acids to form proteins.

- **Lightning** also fixes nitrogen by combining nitrogen and oxygen to form **oxides**. The oxides dissolve in rain and form nitrates in the soil.

- **Denitrifying bacteria** convert nitrates back into nitrogen gas.

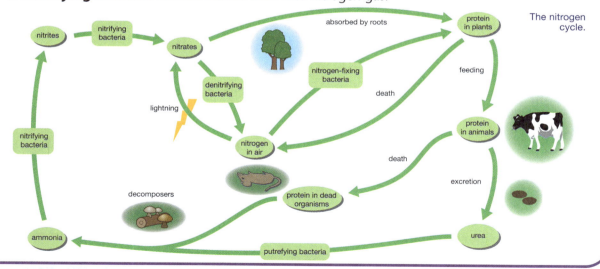

The nitrogen cycle.

Questions

1 Name the process that removes carbon from the atmosphere.

2 Explain how carbon is released from limestone.

3 Name the type of organism that converts protein into nitrates.

4 Describe one way in which nitrogen can be 'fixed'.

B4 Summary

Making food

Many plant cells contain **green chloroplasts**. Chloroplasts contain **chlorophyll**.

Chemical reactions in chloroplasts make **sugars**.

Photosynthesis takes place in **chloroplasts**.

Photosynthesis uses **carbon dioxide** and **water**.

Carbon dioxide enters through **stomata** in **leaves**. **Oxygen** is released in photosynthesis.

Using water

Water can move in and out of cells.

Water inside cells keeps their shape.

Water can enter or leave by **diffusion** or **osmosis**.

Osmosis only works through a partially permeable membrane.

A plant **wilts** when it loses too much water.

Water enters plants through root hairs.

Phloem cells **carry dissolved food** to **roots**.

Xylem cells **carry water** up **stems**.

Water loss from plants is called **transpiration**.

Plants lose water through their **stomata**.

Transferring energy

Energy in plants and animals is transferred in **food chains** and **webs**.

Pyramids of biomass and number show different **trophic levels**.

Food chains rarely exceed five links.

Each trophic level 'loses' energy.

The energy in **biomass** can be used as fuels called **biofuels**, for example to power buses.

Alcohol from fermenting sugar cane can be used as a biofuel.

Fast growing trees can fuel power stations.

Pesticides can build up in food chains.

Intensive farming is very efficient.

Intensive farming uses **insecticides**, **herbicides** and **artificial fertilisers**.

Organic farming does not use manufactured chemicals.

Hydroponics is an example of intensive farming.

Decay and recycling

Carbon and **nitrogen** are recycled naturally.

Carbon dioxide is taken out of the air during **photosynthesis** and is put into the air during burning of fossil fuels and **respiration**.

Nitrogen is put into the air by **denitrifying bacteria** and is removed from the air by **nitrogen-fixing bacteria**.

Bacteria and **fungi** are **decomposers**.

Decomposers cause dead bodies to decay.

The decay of food can be slowed down or stopped by methods of **food preservation**.

Decay is speeded up in warm, damp conditions.

Food preservation methods slow down or stop the growth of bacteria and fungi.

Acids and bases

Neutralising acids

D–C

- An alkali is a **base** which dissolves in water.

- The word equation for **neutralisation** is: acid + base ⟶ salt + water

- Metal oxides and metal hydroxides neutralise acids because they're bases.
  The reaction of a metal oxide or a metal hydroxide with an acid is:
  acid + oxide ⟶ salt + water   acid + hydroxide ⟶ salt + water

- **Carbonates** also neutralise acids to give water and a gas:
  acid + carbonate ⟶ salt + water + carbon dioxide

- A **salt** is made from part of a base and part of an acid.

- To work out the name of a salt, look at the acid and base it was made from.
  The first part of the salt name is from the base and the second part from the acid.

Top Tip!

Nitrates come from nitric acid, chlorides come from hydrochloric acid, sulphates come from sulphuric acid.

> When sodium hydroxide reacts with hydrochloric acid,
> the salt formed is sodium chloride:
> **sodium**   **chloride**
> from the base   from the acid
>
> Na OH + H Cl → NaCl (a salt) / H$_2$O (water)
> base   acid

- The following are the **balanced symbol equations** for the reactions between
  some common acids and bases.

B–A*

$$HCl + NaOH \longrightarrow NaCl + H_2O$$
$$H_2SO_4 + CaCO_3 \longrightarrow CaSO_4 + H_2O + CO_2$$

$$2HNO_3 + CuO \longrightarrow Cu[NO3]_2 + H_2O$$
$$2HCl + CaCO_3 \longrightarrow CaCl_2 + H_2O + CO_2$$

Hydrogen ions

B–A*

- An **acid solution** contains **hydrogen ions, H$^+$** which are responsible for the reactions of an acid.

- An **alkali solution** contains **hydroxide ions, OH$^-$**.

- When an acid **neutralises** an alkali, the hydrogen ions react with the hydroxide ions to make water.
  $$H^+ + OH^- \longrightarrow H_2O$$

The pH scale

D–C

- When an acid is added to alkali, or the other way round, a change in pH happens.

adding an alkali to an acid	adding an acid to an alkali
the pH at the start is low	the pH at the start is high
the pH rises as the alkali neutralises the acid when neutral, the pH = 7	the pH falls as the acid neutralises the alkali when neutral, the pH = 7
when more alkali is added, the pH rises above 7	when more acid is added the pH falls below 7

- **Universal indicator solution** can be used to measure the acidity of a solution.
  A few drops are added to the test solution and then the colour of the solution is compared
  to a standard colour chart. When acid is added to alkali, they neutralise each other.

Questions

Grades D-C

1. What's the base needed to make zinc sulphate?
2. Which salt is made when magnesium hydroxide reacts with sulphuric acid?

Grades B-A*

3. Write a balanced symbol equation for the reaction between HCl and ZnO.
4. Which ions are responsible for making alkaline solutions?

Reacting masses

Relative formula mass

D–C

- **Relative formula masses** need to be added up in the right order if there are brackets in the formula.

B–A*

- When chemicals react, the atoms of the reactants swap places to make new compounds – the products. These products are made from just the same atoms as before. There are the same number of atoms at the end as there were at the start, so the overall mass stays the same.

1 Work out the inside of the bracket first.
16 + 1 = 17

2 Now multiply the bracket by 3.
17 x 3 = 51

Al(OH)$_3$

3 Work out the outside of the bracket.
= 27

4 Finally, add them all together.
27 + 51 = 78

Percentage yield

D–C

- Calculations can be made of how much product is produced in a reaction without knowing the equations for the reactions.

- To calculate **percentage yield**, the following two things must be known:
  – the amount of product made, the 'actual yield'
  – the amount of product that should have been made, the 'predicted yield'.

$$\text{percentage yield} = \frac{\text{actual yield}}{\text{predicted yield}} \times 100$$

B–A*

- Predicted yield can be made by looking at the equation for a reaction.
$ZnCO_3 \longrightarrow ZnO + CO_2$

> To find out how much CO_2 is made when 625 g of $ZnCO_3$ decomposes, the first step is to find the relative formula mass of $ZnCO_3$.
> 65 + 12 + 16 + 16 + 16 = 125
> Then find the relative formula mass of CO_2.
> 12 + 16 + 16 = 44
> The last step is to work out the amount of product made using ratios.
> So if 125 g $ZnCO_3$ gives 44 g CO_2
> then 625 g $ZnCO_3$ gives $\frac{625}{125} \times 44$
> $\qquad\qquad = 220$ g CO_2

Questions

You may need to use the periodic table on page 238 to help you to answer the questions.

Grades D-C

1 Write down the relative formula mass of calcium nitrate, Ca(NO$_3$)$_2$.

Grades B-A*

2 Explain why mass is conserved in a chemical reaction.

Grades D-C

3 Tim made 24 g of crystals instead of 32 g. What's the percentage yield?

Grades B-A*

4 How much carbon dioxide is made in the complete thermal decomposition of 59.5 g nickel carbonate, NiCO$_3$?

Fertilisers and crop yield

Using fertilisers

- Farmers use **fertilisers** to increase their **crop yields**. They must first be dissolved in water before they can be absorbed by plants through their roots.

- Fertilisers increase crop yield by replacing the essential elements used by a previous crop. More nitrogen gets incorporated into plant protein, so there's increased growth.

Relative formula mass

- To calculate the yield when making a fertiliser, you need to calculate its **relative formula mass**. The relative formula mass of ammonium nitrate, NH_4NO_3, is 80.

NH_4NO_3

14   4 x 1 = 4   14   3 x 16 = 48

14 + 4 + 14 + 48 = 80

- Farmers can use relative formula masses to find the percentage of each element in a fertiliser – it's printed on the bag.

$$\text{percentage of element} = \frac{\text{mass of the element in the formula}}{\text{relative formula mass}} \times 100$$

Making fertilisers

- Many fertilisers are **salts**, so they can be made by reacting acids with bases.

acid + base ⟶ salt + water

nitric acid + potassium hydroxide ⟶ potassium nitrate + water
nitric acid + ammonium hydroxide ⟶ ammonium nitrate + water
sulphuric acid + ammonium hydroxide ⟶ ammonium sulphate + water
phosphoric acid + ammonium hydroxide ⟶ ammonium phosphate + water

- In the laboratory, sulphuric acid (acid) is reacted with ammonium hydroxide (base). The amounts used in the reaction must be exactly right, so a **titration** is carried out before mixing the main batch of chemicals.
  – Titrate the alkali with the acid, using an **indicator**.
  – Repeat the titration until three consistent results are obtained.
  – Use the titration result to mix the correct amounts of acid and alkali, without the indicator.
  – The fertiliser made is dissolved in water, so evaporate most of the water using a hot water bath.
  – Leave the remaining solution to crystallise, then filter off the crystals.

The dangers of fertilisers

- Fertilisers must be applied carefully. Rain water dissolves fertilisers, which run off into nearby water courses. If nitrate or phosphate levels in a water course are too high, **eutrophication** occurs.

Eutrophication.

Questions

1 What's the relative formula mass of ammonium phosphate, $(NH4)_3PO_4$?

2 Calculate the percentage of nitrogen in ammonium phosphate.

3 Which acid and base react to make potassium phosphate?

4 Suggest how a solid sample of potassium phosphate could be made. Outline all the main stages.

The Haber process

The Haber process

D–C

- The Haber process uses:
  - an iron catalyst
  - high pressure
  - a temperature of 450 °C
  - a recycling system for unreacted nitrogen and hydrogen.

- The word equation for the Haber process is:
  nitrogen + hydrogen ⇌ ammonia

The Haber process.

- The symbol equation for the Haber process is:
  $N_2 + 3H_2 \rightleftharpoons 2NH_3$

B–A*

- As the reaction is reversible, the **percentage yield** for the reaction can't be 100%.
  - A higher pressure **increases** the percentage yield but high pressures costs more.
  - The high temperature **decreases** the percentage yield. However, higher temperatures do make the reaction go faster.
  - 450 °C is an **optimum temperature** – the yield isn't as good, but that yield is made faster, so a satisfactory amount is produced in the right time.
  - Catalysts don't affect the yield – they just make the reaction go faster.

The costs of ammonia production

D–C

- Different factors affect the cost of making a new substance:
  - **labour** – chemical plants are heavily automated and need few people to operate them
  - **reactants** – hydrogen is made from natural gas or by cracking oil, which costs money; nitrogen has to be cleaned, dried and compressed
  - **recycling** of unreacted materials – means that money isn't wasted
  - **high pressure** – this makes the reaction work better but costs more
  - **energy** – the higher the temperature, the more fuel is needed
  - **reaction rate** – the faster the reaction, the more product is made from the same equipment, so the cheaper it is
  - **pollution control** – reducing pollution is expensive.

B–A*

- Chemical plants work at the conditions that produce the highest percentage yield for a reaction, most cheaply. These are known as the **optimum conditions**.
  - High temperature means higher rates but lower yields, so it runs at the optimum temperature of 450 °C. This temperature means higher energy costs and also lower yields, but the increase in **rate** compensates. The plant produces more ammonia in a day at this temperature than it would at lower temperatures.
  - Total energy costs are not only due to heating costs. The plant needs compressors and pumps to achieve a high pressure. High pressure costs more, so a lower, optimum pressure is used.
  - Although the reaction has a low percentage yield, the unreacted chemicals are recycled, and can go back into the reaction vessel, saving costs.

Questions

(Grades D-C)

1 At what temperature is the Haber process carried out?

(Grades B-A*)

2 Why is this temperature chosen?

(Grades D-C)

3 Why is a low pressure not used in the Haber process?

(Grades B-A*)

4 High pressure gives the highest yields in the Haber process, but it's not used. Explain why.

Detergents

Detergents

- A detergent can be made by **neutralising** an organic acid using an alkali.

  acid + alkali ⟶ salt + water

  It's suitable for cleaning uses because:
  – it dissolves grease stains   – it dissolves in water at the same time.

- New washing powders allow clothes to be washed at low temperatures. It's good for the environment to wash clothes at 40 °C instead of at high temperatures because washing machines have to heat up a lot of water. This needs energy, so the lower the temperature of the water, the less energy is used and the less greenhouse gases are put into the atmosphere.

- Washing clothes at low temperatures is also good for coloured clothes as many dyes are easily damaged by high temperatures. It also means that many more fabrics can be machine washed as their structure would be damaged at higher temperatures.

- A detergent molecule is made of two parts:
  – the **hydrophilic** part – forms bonds with the water and pulls the grease off the fabric or dish
  – the **hydrophobic** part – forms bonds with the oil or grease.

hydrophobic end, dissolves in grease   hydrophilic end, dissolves in water

A detergent molecule.

How a detergent works.

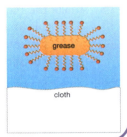

Dry cleaning

- Some fabrics will be damaged if they are washed in water, so they must be **dry-cleaned**. A dry-cleaning machine washes clothes in an organic solvent. The word 'dry' doesn't mean that no liquids are used, just that the liquid solvent isn't water.

- Most of the stains on clothing contain grease from the skin or from food. Grease-based stains won't dissolve in water, but they will dissolve easily in a dry-cleaning solvent.

- Forces between molecules are called **intermolecular molecules**.
  – These forces hold molecules of grease together and molecules of dry-cleaning solvent together.
  – The forces join anything to anything, so dry-cleaning solvent molecules also bind to grease. The grease then dissolves in the solvent.
  – Molecules of water are held together by stronger intermolecular forces called **hydrogen bonds**. The water molecules can't stick to the grease because they're sticking to each other much too strongly.

grease sticks to grease

molecules of dry-cleaning solvent stick to other molecules of dry-cleaning solvent

molecules of dry-cleaning solvent stick to grease

water sticks strongly to water

water sticks to water too strongly to stick to grease

Questions

Grades D-C

1 What are the two reactants used in making a detergent?

Grades B-A*

2 What does the hydrophobic tail of the detergent dissolve?

Grades D-C

3 Give two reasons why a dry-cleaning method may be used.

Grades B-A*

4 Which intermolecular forces are stronger: grease to water or water to water?

Batch or continuous?

Batch and continuous manufacturing

D–C

- Drugs companies make medicines in small batches, which are then stored.
  - New batches are made when the stored medicine runs low. If a lot of one medicine is needed, several batches can be made at the same time.
  - Once they have made a batch of one drug, it's easy to switch to making a different drug.

- The large scale production of **ammonia** is different to the small scale production of pharmaceuticals as it's a continuous process.

carry out the reaction

put reagents into the reaction vessel

BATCH PROCESS

filter off the product

clean out the reaction vessel and find out what needs to be made for the next batch

B–A*

- A **continuous** process **plant** is effective because it works at full capacity all the time. It costs an enormous amount to build, but once running it makes a large amount of product and employs very few people, making the cost per tonne very small. A disadvantage is that the reaction vessels and pipes are only designed to work well at one level of output. What they make or how much can't easily be changed.

- **Batch processes**, by comparison, are flexible. It's easy to change from making one compound to another. Each batch has to be supervised, so labour costs are higher. Also, time spent filling and emptying reaction vessels means the vessels aren't producing chemicals, so they're not used as efficiently as in a continuous process.

Why are medicines so expensive?

D–C

- The high costs of making and developing medicines and pharmaceutical drugs include:
  - **strict safety laws**
  - **research and development** – many take years to develop
  - **raw materials** – may be rare and costly
  - **labour intensive** – because medicines are made by a batch process, less automation can be used.

Drug development

B–A*

- Whether or not a drug is developed depends on a number of economic considerations.
  - Drug development is expensive because it takes time and labour costs are high.
  - Testing drugs to meet legal requirements can take years.
  - There must be enough demand for the drug.
  - The length of time taken to pay back the initial investment is long.

Questions

(Grades D-C)

1 What's a batch process?

(Grades B-A*)

2 Write down one advantage of using a continuous process.

(Grades D-C)

3 Give two reasons why medicines are expensive to develop.

(Grades B-A*)

4 How long can it take to research, develop, test and patent a new drug?

Nanochemistry

Forms of carbon

- There are three forms of **carbon** shown in the table.

	diamond	graphite	buckminster fullerene
use	cutting tools – very hard jewellery – lustrous and colourless	pencil leads – slippery lubricants – slippery electrode – conducts electricity and has high melting point	semiconductors in electrical circuits (nanotubes) industrial catalysts reinforce graphite in tennis rackets (nanotubes)
structure			
properties	doesn't conduct electricity because it has no free electrons hard with a high melting point because of the presence of many strong covalent bonds	conducts electricity because of delocalised electrons that can move slippery because layers of carbon atoms are weakly held together and can slide easily over each other high melting point because there are many strong covalent bonds to break	nanotubes have a large surface area so they can be used as cages to trap or transport other molecules or as carriers for catalysts

Allotropes

- Different forms of the same element are called **allotropes**. Diamond, graphite and the fullerenes are all **allotropes of carbon**.

- The different arrangements of atoms give each allotrope different properties. In diamond, each atom is held by **covalent bonds** to four other atoms, **tetrahedrally**, which are bonded further in different directions. This is called a **giant structure**. So many strong covalent bonds make the diamond hard and very difficult to melt. The bonding results in no free electrons, so it doesn't conduct electricity.

Nano properties

- A **fullerene** changes at the **nanoscale**. The shape of the individual particles – balls or tubes, sieves or cages – is their nanostructure. This gives them their **nano properties**. These are different from **bulk properties**, which are the properties of large amounts of a material.

- Nanoparticles are made in a different way. One type of nanoparticle is used and then bits are knocked off or stuck on. Structural engineering happens at a molecular level.

Questions

Grades D-C

1 Which properties of graphite make it useful as an electrode?
2 Write down two potential uses of nanotubes.

Grades B-A*

3 Explain why graphite can conduct electricity but diamond can't.
4 Explain one use of a fullerene as a cage.

How pure is our water?

Water purification

- The water in a river is cloudy and often not fit to drink. It may also contain pollutants such as nitrates from fertiliser run off, lead compounds from lead pipes, and pesticides from spraying near to water resources. To make clean drinking water it's passed through a **water purification** works.

A sedimentation tank at a water treatment works.

- There are three main stages in water purification:
  - **sedimentation** of particles – larger bits drop to the bottom
  - **filtration** of very fine particles – sand is used to filter out finer particles
  - **chlorination** – kills microbes.

- Some soluble substances remain in the water. Some of these can be poisonous, for example pesticides and nitrates. Extra processes are needed to treat these.

- Seawater has many substances dissolved in it so it's undrinkable. Techniques such as **distillation** must be used to remove the dissolved substances. Distillation uses huge amounts of energy, and is very expensive. It's only used when there's no fresh water.

The importance of clean water

- Clean water saves more lives than medicines. That's why, after disasters and in developing countries, relief organisations concentrate on providing clean water supplies.

- Water is a **renewable resource**, but that doesn't mean the supply is endless. If there isn't enough rain in the winter, reservoirs don't fill up properly for the rest of the year.

- Producing tap water does incur costs. It takes energy to pump and to purify it – all of which increases climate change.

Water tests

- In a precipitation reaction, two solutions react to form a solid that doesn't dissolve.

  lead nitrate + sodium **sulphate** $\longrightarrow$ lead sulphate (**white** precipitate) + sodium nitrate
  silver nitrate + sodium **chloride** $\longrightarrow$ silver chloride (**white** precipitate) + sodium nitrate
  silver nitrate + sodium **bromide** $\longrightarrow$ silver bromide (**cream** precipitate) + sodium nitrate
  silver nitrate + sodium **iodide** $\longrightarrow$ silver iodide (**yellow** precipitate) + sodium nitrate

- The **balanced symbol equations** for the precipitation reactions above are:

  $Pb(NO_3)_2(aq) + Na_2SO_4(aq) \longrightarrow PbSO_4(s) + 2NaNo_3(aq)$
  $AgNO_3(aq) + NaCl(aq) \longrightarrow AgCl(s) + NaNO_3(aq)$
  $AgNO_3(aq) + NaBr(aq) \longrightarrow AgBr(s) + NaNO_3(aq)$
  $AgNO_3(aq) + NaI(aq) \longrightarrow AgI(s) + NaNO_3(aq)$

Questions

Grades D-C

1 Explain why filtration is used in the water purification process.

Grades B-A*

2 Explain why distillation uses large amounts of energy.

Grades D-C

3 What type of reaction takes place between barium chloride and sulphates?

Grades B-A*

4 Write a balanced symbol equation for the reaction between silver nitrate solution and magnesium chloride, $MgBr_2$.

C4 Summary

Nitrogen and hydrogen make ammonia in the **Haber process**.
$N_2 + 3H_2 \rightleftharpoons 2NH_3$

Ammonia is made all the time in a **continuous process**. Pharmaceutical drugs are made on a smaller scale by a **batch process**.

We can change the **conditions** in the Haber process to give us the best **yield**.
– High pressure increases yield.
– High temperature decreases yield but increases rate.
– Optimum temperatures are used.

Ammonia reacts as a **base** to form fertilisers such as ammonium nitrate from nitric acid, and ammonium phosphate from phosphoric acid.

Chemical industry

If we know the mass of the reactants, we can work out what mass of products to expect.

Different factors affect the **cost** of making fertilisers or any new substance. High pressures mean higher energy costs.

Industry makes chemicals such as **fertilisers**. They must be cheap enough to use.

We can measure the **percentage yield** of a reaction.
$$\% \text{ yield} = \frac{\text{actual yield}}{\text{predicted yield}} \times 100$$

Water resources are found in lakes, rivers, aquifers and reservoirs. Water needs testing and purifying before use. Purification includes:
– filtration
– sedimentation
– chlorination.

Fertilisers make crops grow bigger as they provide plants with extra nitrogen, phosphorus and potassium. These are essential chemical elements for plant growth.

Detergents are **molecules** that combine with both grease and water. They have a hydrophilic head and a hydrophobic tail. Washing powders contain detergents but some also contain enzymes, which allow clothes to be washed at a lower temperature. This saves energy.

Diamonds are used in cutting tools and jewellery. They're **very hard** and have a **high melting point**.

Nanochemistry

Graphite is carbon. It's **slippery** so it can be used as a lubricant. It also **conducts electricity** so it can be used as electrodes.

The element carbon can exist in different forms. This is due to differences at the **nanoscale**.

Fullerenes were discovered fairly recently. Buckminster fullerene has the formula C_{60}.

Sparks!

Electrons

D–C

- An **atom** consists of a small positively charged nucleus surrounded by an equal number of negatively charged electrons.

- In a stable, neutral atom, there are the same amounts of **positive** and **negative** charges.

- All **electrostatic** effects are due to the movement of electrons.

- The law of electric charge states that: like charges repel, unlike charges attract.

> **Top Tip!**
>
> It's only electrons that move in an atom.

electron
proton
neutron

The charges in a neutral atom balance. The atom has four electrons. How many protons does it have?

B–A*

- When a polythene rod is rubbed with a duster, electrons are transferred from the duster to the polythene, making the polythene rod negatively charged.

- When an acetate rod is rubbed with a duster, electrons are transferred from the acetate to the duster, leaving the acetate rod positively charged.

- In general an object has:
  - a **negative** charge due to an **excess of electrons**
  - a **positive** charge due to a **lack of electrons**.

duster
e–
e–
e–
polythene rod

Why does the polythene rod become negatively charged when it's rubbed with a duster?

Electric shocks

D–C

- When inflammable gases or vapours are present, or there is a high concentration of oxygen, a spark from static electricity could ignite the gases or vapours and cause an explosion.

- If a person touches something at a high **voltage**, large amounts of electric charge may flow through their body to earth.

- **Current** is the rate of flow of charge. Even small currents can be fatal.

- Static electricity can be a nuisance but not dangerous.
  - Dust and dirt are attracted to insulators, such as television screens.
  - Clothes made from synthetic materials often 'cling' to each other and to the body.

B–A*

- Electric shocks can be avoided in the following ways:
  - If an object that is likely to become charged is connected to earth, any build up of charge would immediately flow down the earth wire.
  - In a factory where machinery is at risk of becoming charged, the operators stand on insulating rubber mats so that charge cannot flow through them to earth.
  - Shoes with insulating soles are worn by workers if there is a risk of charge building up so that charge cannot flow through them to earth.

- Anti-static sprays, liquids and cloths made from conducting materials carry away electric charge. This prevents a build up of charge.

Questions

Grades D-C

1 What happens when **a** two acetate rods are brought near each other? **b** a polythene and an acetate rod are brought near each other?

Grades B-A*

2 Why does an acetate rod become positively charged when rubbed with a duster?

Grades D-C

3 Why must a mobile phone not be used on a petrol station forecourt?

Grades B-A*

4 Suggest how tumble drier sheets can reduce static 'cling' in your clothes.

Uses of electrostatics

Defibrillators

- Defibrillation is a procedure to restore a regular **heart rhythm** by delivering an electric shock through the chest wall to the heart.
  - Two **paddles** are charged from a high voltage supply.
  - They are then placed firmly on the patient's chest to ensure good electrical contact.
  - Electric charge is passed through the patient to make their heart contract.
  - Great care is taken to ensure that the operator does not receive an electric shock.

D–C

- A shock from a defibrillator supplies about 400 J of energy in a few milliseconds.

  If a defibrillator is switched on for 5 milliseconds (0.005 s), the power can be calculated from:

  $$\text{power} = \frac{\text{energy}}{\text{time}} = \frac{400}{0.005} = 80\ 000\ \text{W}$$

B–A*

Paint sprayers

- Static electricity is used in paint spraying.
  - The **spray gun** is charged.
  - All the paint particles become charged with the same charge.
  - Like charges **repel**, so the paint particles spread out giving a fine spray.
  - The object to be painted is given the *opposite* charge to the paint.
  - Opposite charges **attract**, so the paint is attracted to the object and sticks to it.
  - The object gets an even coat, with limited paint wasted.

D–C

How an electrostatic paint sprayer works.

nozzle is charged up positively

object to be painted is negatively charged

- If the object to be painted is *not* charged, the paint moves onto it but:
  - the object becomes charged from the paint, gaining the *same* charge
  - further paint droplets are repelled away from the object.

B–A*

Dust precipitators

- A dust precipitator removes harmful particles from the chimneys of factories and power stations that pollute the atmosphere.

D–C

  - A metal grid (or wires) is placed in the chimney and given a large negative charge from a high-voltage supply.
  - As the soot particles pass close to the wires, the soot particles become negatively charged.

clean smoke released into air

wires

plates

soot particles

end-on view of wires

soot particles sticking to plate

How a dust precipitator works.

B–A*

  - Like charges repel, so the soot particles are repelled away from the wires. They are attracted to the positively charged plates and stick to them.
  - When the particles get big enough, they fall back down the chimney.

Questions

Grades D-C

1 Suggest why the paddles of a defibrillator must be placed firmly on the bare chest of the patient.

Grades B-A*

2 A defibrillator has a power rating of 100 000 W. For how long must it be switched on to provide 500 J of energy?

Grades D-C

3 Ali is spray painting the frame of his bicycle. Explain why he will get a better finish if he earths the frame.

Grades B-A*

4 Why do the wires in an electrostatic dust precipitator need to be at a high voltage?

Safe electricals

Electric circuits

- An electric **current** is a flow of electric **charge**. Charge is carried by negatively charged **electrons**. The electrons flow in the *opposite* direction to the conventional current.

- The current is measured in **amperes** (A) using an **ammeter** connected in **series**.

- The current in a circuit is due to electron movement.
  - Electrons are pushed around the circuit by the battery.
  - They bump into atoms in the resistor giving them energy.
  - Their energy makes the atoms vibrate more, so the resistor gets hotter and its resistance increases.

Resistors and resistance

This circuit has a variable resistor and acts as a dimmer switch.

- A **variable resistor** (or **rheostat**) changes the resistance.

- The **potential difference** (pd) between two points in a circuit is the difference in **voltage** between the two points.

- Potential difference is measured in **volts** (V) using a **voltmeter** connected in **parallel**.
  - For a fixed resistor, as the potential difference across it **increases**, the **current increases**.
  - For a fixed power supply, as the **resistance increases**, the **current decreases**.

- The formula for resistance is: $resistance = \dfrac{potential\ difference}{current}$ $R = \dfrac{V}{I}$
  - Resistance is measured in **ohms** (Ω).

  If the pd across a lamp is 5.0 V when the current is 0.2 A, the resistance is: $R = \dfrac{V}{I} = \dfrac{5.0}{0.2} = 25\ \Omega$

- The formula for resistance can be rearranged to find out:
  - potential difference $V = IR$ – current $I = \dfrac{V}{R}$

Top Tip!

Always remember to include the correct unit.

Live, neutral and earth wires

- The **live wire** carries a high voltage around the house.

- The **neutral wire** completes the circuit, providing a return path for the current.

- The **earth wire** is connected to the case of an appliance to prevent it becoming live.

Insulation and fuses

- A fuse contains wire which melts, breaking the circuit, if the current becomes too large.

- No current can flow, preventing overheating and further damage to the appliance.

- Earth wires and fuses stop a person receiving an electric shock if they touch a faulty appliance. As soon as the case becomes 'live', a large current flows in the earth and live wires and fuse 'blows'.

- A re-settable fuse (circuit-breaker) doesn't need to be replaced to restore power; it can be re-set.

Questions

(Grades D-C)

1 The pd across a resistor is 4.0 V when the current through it is 0.5 A. What is its resistance?

(Grades B-A*)

2 Calculate the current in a 10 Ω resistor when the pd across it is 4 V.

(Grades D-C)

3 Explain how a fuse protects an appliance.

(Grades B-A*)

4 How does an earth wire stop a person receiving an electric shock?

Ultrasound

Longitudinal waves

- All **sound**, including **ultrasound**, is produced by **vibrating** particles that form a **longitudinal wave**.

- The features of longitudinal sound waves are:
  - They can't travel through a vacuum. The denser the medium, the faster a wave travels.
  - The higher the **frequency** or **pitch**, the smaller the **wavelength**.
  - The **louder** the sound, or the more powerful the ultrasound, the more energy is carried by the wave and the larger its amplitude.

- **Ultrasound** is sound of a higher **frequency** than humans can hear. It travels as a pressure wave – **compressions** and **rarefactions**.

- In a longitudinal wave the vibrations of the particles are in the same direction as the wave.

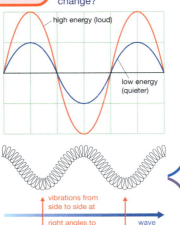

Comparing a loud sound with a quieter sound. How does the amplitude change?

high energy (loud)

low energy (quieter)

vibrations in same direction as wave

wave direction

A longitudinal wave.
A transverse wave.

vibrations from side to side at right angles to direction of wave

wave direction

D–C

B–A*

- In a transverse wave the vibrations of the particles are at right angles to the direction of the wave.

Uses of ultrasound

- When **ultrasound** is used to break down kidney stones:
  - a high-powered ultrasound beam is directed at the kidney stones
  - the ultrasound energy breaks the stones down into smaller pieces
  - the tiny pieces are then excreted from the body in the normal way.

- When **ultrasound** is used in a **body scan**, a **pulse** of ultrasound is sent into body.
  - At each boundary between different tissues some ultrasound is **reflected** and the rest is transmitted.
  - The returning **echoes** are recorded and used to build up an image of the internal structure.

- Ultrasound can be used for **body scans** because:
  - When ultrasound is reflected from different **interfaces** in the body, the depth of each structure is calculated by using the formula distance = speed × time, knowing the speed of ultrasound for different tissue types and the time fo the echo to return.
  - The proportion of ultrasound reflected at each interface depends on the densities of the adjoining tissues and the speed of sound in the adjoining tissues.
  - If the tissues are very different (e.g. blood and bone) most of the ultrasound is reflected, leaving very little to penetrate further into the body.
  - The information gained is used to produce an image of the part of the body scanned.

- Ultrasound is preferred to X-rays because:
  - it differentiates between soft tissues better
  - it doesn't damage living cells.

D–C

B–A*

abdominal wall   organ   vertebra

ultrasound probe

double reflection

How a body scan works.   reflections from boundaries   path of ultrasound

Questions

1 Body fat is denser than air. In which medium will ultrasound travel faster?

2 Is a Mexican wave a longitudinal or a transverse wave?

3 An ultrasound pulse travels 20 cm further when it is reflected from one side of the head of a foetus compared with the other side. How big is the head?

4 The time delay for an ultrasound echo in a soft tissue was 0.0004 seconds. At what depth was it reflected if the speed of ultrasound in the soft tissue is 1200 m/s?

Treatment

Using radiation

D–C

- Radiation emitted from the nucleus of an unstable atom can be **alpha** (α), **beta** (β) or **gamma** (γ).
  - **Alpha radiation** is absorbed by the skin so is of no use for diagnosis or therapy.
  - **Beta radiation** passes through skin, but not bone. Its medical applications are limited but it's used, for example, to treat the eyes.
  - **Gamma radiation** is very penetrating and is used in medicine. **Cobalt-60** is a gamma-emitting radioactive material that is widely used to treat cancers.

- When nuclear radiation passes through a material it causes **ionisation**. Ionising radiation damages living cells, increasing the risk of **cancer**.

- Cancer cells within the body can be destroyed by exposing the affected area to large amounts of radiation. This is called **radiotherapy**.

Comparing X-rays and gamma rays

D–C

- Gamma rays and X-rays have similar wavelengths but are produced in different ways.

An X-ray tube to produce X-rays.

B–A*

- **X-rays** are made by firing high-speed **electrons** at metal targets.

- An X-ray machine allows the rate of production and energy of the X-rays to be controlled, but you can't change the gamma radiation emitted from a particular radioactive source.

- When the **nucleus** of an atom of a radioactive substance decays, it emits an alpha or a beta particle and loses any surplus **energy** by emitting **gamma rays**.

Tracers

D–C

- A radioactive **tracer** is used to investigate inside a patient's body without surgery.
  - Technetium-99m is a commonly used tracer. It emits only gamma radiation.
  - Iodine-123 emits gamma radiation. It's used as a tracer to investigate the thyroid gland.

B–A*

- The radioactive tracer being used is mixed with food or drink or injected into the body.
  - Its progress through the body is monitored using a detector such as a **gamma camera** connected to a computer.

Treating cancer

Treating a brain tumour with three sources of gamma radiation.

B–A*

- A radioisotope is used to destroy a tumour in the body.

- Three sources of radiation, each providing one-third of the required dose, are arranged around the patient with the tumour at the centre.
  - The healthy tissue only receives one-third of the dose, which limits damage to healthy tissue.

- Or each radiation source is slowly rotated around the patient. The tumour receives constant radiation but healthy tissue receives only intermittent doses.

Questions

Grades D-C

1 What is 'radiotherapy'?

Grades B-A*

2 Give two similarities and two differences between X-rays and gamma rays.

Grades D-C

3 Why is iodine-123 used as a tracer in medicine?

Grades B-A*

4 Explain why gamma radiation is often directed at a tumour from several different directions.

What is radioactivity?

Radioactive decay

- **Radioactive** substances decay naturally, giving out **alpha**, **beta** and/or **gamma** radiation.

- Radioactive decay is a **random** process; it isn't possible to predict exactly when a nucleus will decay.
  - There are so many atoms in even the smallest amount of **radioisotope** that the average count rate will always be about the same.

- Radioisotopes have **unstable nuclei**. Their nuclear particles aren't held together strongly enough.

- The **half-life** of a radioisotope is the average time for half the nuclei present to decay. The half-life cannot be changed.

D–C
B–A*

The nucleus

- A **nucleon** is a particle found in the nucleus. So, **protons** and **neutrons** are nucleons.

- The nucleus of an atom can be represented as:

  $$^A_Z X$$ where
  - A = atomic mass (or nucleon number)
  - Z = atomic number (or proton number)
  - X = chemical symbol for the element.

  Z = the number of protons in the nucleus, so the number of neutrons = $(A - Z)$.

  The carbon isotope $^{14}_6 C$ has 6 protons and 8 neutrons in its nucleus.

Top Tip!
Nucleons cannot be lost. Charge is always conserved.

Top Tip!
Know the difference between alpha, beta and gamma radiation.

B–A*

What are alpha and beta particles?

- When an **alpha** or a **beta** particle is emitted from the nucleus of an atom, the remaining nucleus is a **different** element.

D–C

	alpha (α) particle	beta (β) particle
properties	positively charged has a large mass is a helium nucleus has helium gas around it consists of 2 protons and 2 neutrons	negatively charged has a very small mass travels very fast is an electron
during decay	mass number decreases by 4 nucleus has two less neutrons nucleus has two less protons atomic number decreases by 2	mass number is unchanged nucleus has one less neutron nucleus has one more proton atomic number increases by one
nuclear equation for decay	$^{238}_{92}U \rightarrow {}^{234}_{90}Th + {}^4_2He$ (α-particle)	$^{14}_6C \rightarrow {}^{14}_7N + {}^0_{-1}e$ (β-particle)

B–A*

alpha decay — α-particle
uranium-238 thorium-234 γ-ray

beta decay — β-particle
carbon-14 nitrogen-14

Questions

Grades D-C

1. Three successive measurements of the activity of a radioactive source are 510 Bq, 495 Bq and 523 Bq. Why are they different?

3. Complete the nuclear equation:
   $^{219}_{86}Rn \rightarrow \underline{\quad}Po + {}^4_2He$

Grades B-A*

2. The activity of a radioactive sample took 8 hours to decrease from 5000 Bq to 1250 Bq. What is its half-life?

Uses of radioisotopes

Background radiation

D–C

- **Background radiation** is due to:
  – radioactive substances present in **rocks** (especially **granite**) and soil
  – **cosmic rays** from Space.

B–A*

- Most **background radiation** is from natural sources but some comes from human activity. This is shown in the pie chart.

nuclear power — nuclear weapons testing
medical
radioactivity in the air
food
rocks
radiation from Space (cosmic rays)

Top Tip!

Remember to allow for background radiation when measuring the activity of a source of radiation.

Tracers

D–C

- In addition to medical applications (see page 107), **tracers** are used to:
  – track the dispersal of waste materials
  – find leaks or blockages in underground pipes
  – track the route of underground pipes.

B–A*

- When using a tracer to locate a leak in an underground pipe:
  – a very small amount of a gamma emitter is put into the pipe
  – a detector is passed along the ground above the path of the pipe
  – an increase in activity is detected in the region of the leak and little or no activity is detected after this point.

Geiger-Müller tube detects increased level of radiation

pipe   leak

radioisotope introduced into pipe, emits gamma radiation

Dating rocks

D–C

- Some **rock** types such as granite contain traces of **uranium**, a radioactive material.
  – The **uranium** isotopes present in the rocks go through a series of decays, eventually forming a stable isotope of **lead**.
  – By comparing the amounts of uranium and lead present in a rock sample, its approximate age can be found.

B–A*

- **Uranium-238** decays, with a very long half-life of 4500 million years.
  – $^{238}_{92}U \rightarrow \ ^{234}_{90}Th \ (+ \ ^{4}_{2}He) \rightarrow \ ^{234}_{91}Pa \ (+ \ ^{0}_{-1}e) \rightarrow$ .... $\rightarrow \ ^{206}_{82}Pb$ (stable)

- The proportion of lead increases as time increases. If there are equal quantities of $^{238}_{92}U$ and $^{206}_{82}Pb$, the rock is 4500 million years (one half-life) old.

Radiocarbon-dating

D–C

- **Carbon-14** is a radioactive isotope of carbon that is present in all living things. By measuring the amount of carbon-14 present in an archaeological find, its approximate age can be found.

B–A*

  – When an object dies, no more carbon-14 is produced.
  – As the carbon-14 decays, the activity of the sample decreases.
  – The ratio of current activity from living matter to the sample activity leads to a reasonably accurate date.

Questions

Grades D-C

1 Why can't carbon dating be used to find the age of an iron tool?

Grades B-A*

2 Why must a gamma source be used as a tracer rather than an alpha or beta source?

Grades D-C

3 Why would you expect a new rock to contain a bigger proportion of uranium to lead than an old rock?

Grades B-A*

4 The half life of carbon-14 is 5600 years. Why can't it be used to date an object believed to be about 100 years old?

Fission

Artificial radioactivity

- **Artificial radioisotopes** can be produced by bombarding atoms with the neutrons present in a nuclear reactor.
  - Neutrons are uncharged so they're easily captured by many **nuclei**, producing **unstable isotopes**.

D–C

Nuclear power stations

- Natural uranium consists of two **isotopes**, uranium-235 and uranium-238.
  - The '**enriched uranium**' used as **fuel** in a nuclear power station contains a greater proportion of the uranium-235 isotope than occurs naturally.

- **Fission** occurs when a large unstable **nucleus** is split up and energy is released as **heat**.
  - The heat is used to boil water to produce steam.
  - The pressure of the steam acting on the turbine blades makes it turn.
  - The rotating turbine turns the generator, producing electricity.

- In a nuclear power station, atoms of uranium-235 are bombarded with neutrons. This causes the nucleus to split, releasing energy.

- A typical **fission** can be shown as:

$$^{235}_{92}U + ^{1}_{0}n \rightarrow ^{90}_{36}Kr + ^{143}_{56}Ba + 3(^{1}_{0}n) + \gamma\text{-rays}$$

The extra neutrons emitted cause further uranium nuclei to split. This is described as a **chain reaction** and produces a large amount of energy.

D–C

B–A*

Uranium-235 undergoes a chain reaction that produces a large amount of energy.

a neutron is absorbed by the nucleus of a uranium-235 atom

the nucleus is now less stable than before

it splits into two parts and releases energy

several neutrons are also produced – these may go on to strike the nuclei of other atoms causing further fission reactions

this is called a chain reaction

Radioactive waste

- Nuclear fission produces **radioactive waste** that has to be handled carefully and disposed of safely.
  - **Very low-level waste**, such as that produced by medical applications, is placed in sealed plastic bags then buried or burned under strict controls.
  - Other **low-level waste** may be embedded in glass discs and buried in the sea.
  - **High-level waste**, such as spent fuel rods, is re-processed to make more radioactive materials.

D–C

Controlling nuclear fission

- The output of a nuclear reactor can be controlled.
  - A **graphite moderator** between the fuel rods **slows down** the fast-moving neutrons emitted during fission. Slow-moving neutrons are more likely to be captured by other uranium nuclei.
  - **Boron control rods** can be raised or lowered. Boron **absorbs neutrons**, so fewer neutrons are available to split more uranium nuclei. This controls the rate of fission.

B–A*

boron control rods

steam to turbo-generators

hot gas

graphite moderator

uranium fuel rods

heat exchanger

cold water

gas flow

cold gas

pump

core

thick concrete shield to absorb neutrons and other radiations

Questions

Grades D-C

1 What is meant by 'enriched uranium'?

Grades B-A*

2 How does a chain reaction enable a large amount of energy to be produced in the fission of uranium?

Grades D-C

3 Why is radioactive waste dangerous?

Grades B-A*

4 What do the control rods do in a nuclear reactor?

P4 Summary

Electrostatics

There are two kinds of electric charge, positive and negative.
– **like charges repel**
– **unlike charges attract**.

Electrostatic effects are caused by the transfer of electrons.
– A **positively charged** object **lacks electrons**.
– A **negatively charged** object has an **excess of electrons**.

The chance of getting an electric shock can be reduced by:
– correct earthing
– standing on insulating mats
– wearing shoes with insulating soles.

Some uses of electrostatics are:
– defibrillators
– paint sprayers
– photocopiers.
– dust precipitators.

Using electricity safely

An electric current is a flow of electric charge. The **potential difference** (pd) between two points in a circuit is the difference in voltage between the two points.

$$resistance\ (\Omega) = \frac{pd\ (V)}{current\ (A)}$$

In a plug:
L = live (brown)
N = neutral (blue)
E = earth (green and yellow stripe)
The fuse is connected in the live side. It melts if the current exceeds its stated value, breaking the circuit.

Ultrasound

Medical uses of ultrasound include:
– scans that allow a doctor to see inside you without surgery
– measure the rate of blood flow in the body
– break up kidney or gall stones.

Ultrasound is sound of a higher frequency than we can hear (above 20 000 Hz).

Longitudinal waves – e.g. sound and ultrasound.
Transverse waves – e.g. light.
– **Wavelength** (λ) is the distance occupied by one complete wave.
– **Frequency** (f), measured in hertz (Hz), is the number of complete waves in 1 second.
– **Amplitude** is the maximum distance a particle moves from its normal position.

Nuclear radiation

Medical uses of radioactivity include:
– for diagnosis, as a tracer
– to sterilise equipment
– to treat cancers.
Only β- and γ-radiation can pass through skin. γ-radiation is the most widely used for medical purposes as it is the most penetrating.
Other uses:
– smoke detector
– tracers in industry
– dating rocks and old materials.

Fission is the splitting of a large nucleus, such as uranium, releasing energy. This can set up a chain reaction producing a large amount of energy, as in a nuclear power station or atomic bomb.

Nuclear equations represent α- and β-decay.
$$^{238}_{92}U \rightarrow ^{234}_{90}Th + ^{4}_{2}He\ \alpha\text{-particle})$$
$$^{14}_{92}U \rightarrow ^{14}_{90}Th + ^{0}_{2}He\ (\beta\text{-particle})$$

Nuclear radiation is emitted from the nuclei of radioactive materials:
– α-**particle** – helium nucleus
– β-**particle** – fast-moving electron
– γ-**radiation** – electromagnetic waves with a very short wavelength. X-rays are produced by firing high-speed electrons at a metal target. X-rays and gamma rays are similar but X-rays are easier to control than gamma rays.

Half-life of a radioisotope is the average time for half the nuclei present to decay.

Checklist – Science Unit 1

B1 Understanding ourselves

- I can explain the possible consequences of having high blood pressure.
- I can state the word equation for respiration.
- I know that religion, personal choice and medical issues can influence diet.
- I can calculate BMI and RDA.
- I can describe how white blood cells defend against pathogens.
- I can explain the difference between passive and active immunity.
- I can describe the functions of the main parts of the eye.
- I know the classification and effects of the different types of drugs.
- I can interpret data on the effects of smoking.
- I can describe vasodilation and vasoconstriction.
- I know that sex hormones can be used as contraceptives and for fertility treatment.
- I know that chromosomes are made of DNA.
- I can carry out a genetic cross to predict the possibilities of inherited disorders passing to offspring.
- I know that inherited characteristics can be dominant or recessive.

C1 Carbon chemistry

- I know that cooking food is a chemical change as a new substance is made and it's irreversible.
- I know that protein molecules in eggs and meat change shape when the food is cooked.
- I know that when the shape of a protein changes it's called denaturing.
- I know that emulsifiers are molecules that have a water-loving part and an oil- or fat-loving part.
- I know that alcohols react with acids to make an ester and water.
- I know that a solute is the substance dissolved in a solvent to make a solution that doesn't separate.
- I know that crude oil is a non-renewable fossil fuel, which is a mixture of many hydrocarbons.
- I know that petrol is a crude oil fraction with a low boiling point, which exits at the top of the tower.
- I know that polymerisation is a process which requires high pressure and a catalyst.
- I know that a hydrocarbon is a compound formed between carbon atoms and hydrogen atoms only.
- I know that alkenes are hydrocarbons with one or more double bonds between carbon atoms.
- I know that complete combustion of a hydrocarbon fuel makes carbon dioxide and water only.
- I know that an exothermic reaction is one where energy is released into the surroundings.
- I can work out that: energy transferred = mass of water × 4.2 × temperature change.

P1 Energy for the home

- I can explain the difference between temperature and heat.
- I can explain what is meant by specific heat capacity and specific latent heat.
- I can perform energy calculations involving specific heat capacity and specific latent heat.
- I can explain why different forms of domestic insulation work in terms of energy transfer.
- I can calculate energy efficiency.
- I can explain why infrared radiation is used for cooking.
- I can explain why microwaves are used for cooking and for communication.
- I can explain the advantages of using digital signals instead of analogue signals.
- I can describe total internal reflection and its use in optical fibres, including the endoscope.
- I can explain how wireless signals are used for global communication.
- I can explain why there's sometimes interference with microwave and radio signals.
- I know the properties of a transverse wave and how light is used as a transverse wave.
- I can describe earthquake waves and how they can tell us about the structure of the Earth.
- I know some of the effects of natural events and human activity on weather patterns.

Checklist – Science Unit 2

B2 Understanding our Environment

- I know how to collect and use data to estimate a population. ☐
- I know how to use a key to identify plants and animals. ☐
- I know the characteristics of the different vertebrate groups. ☐
- I can state the word and formula equation for photosynthesis. ☐
- I can describe the effect of increased light, temperature and carbon dioxide on photosynthesis rate. ☐
- I can explain how similar organisms compete for the same ecological niche. ☐
- I can explain how the size of a predator population will affect the prey population. ☐
- I can explain how adaptations of organisms determine their distribution and abundance. ☐
- I can explain how camels, polar bears and cactus are adapted to their habitat. ☐
- I can describe how organisms became fossilised. ☐
- I know that developed countries have a greater effect on world pollution. ☐
- I can explain the effects of increased pollution on climate change, acid rain and the ozone. ☐
- I can describe ways in which animals become extinct. ☐
- I can explain sustainable development and describe how it may protect endangered species. ☐

C2 Rocks and Metals

- I know that paint is a colloid of particles dispersed with particles of a liquid, but not dissolved. ☐
- I know that thermochromic pigments change colour when heated or cooled. ☐
- I know that brick is made from clay, glass from sand and aluminium and iron from ores. ☐
- I know that the decomposition of limestone is: calcium carbonate → calcium oxide + carbon dioxide. ☐
- I know that the outer layer of the Earth is continental plates with oceanic plates under oceans. ☐
- I know that the size of crystals in an igneous rock is related to the rate of cooling of molten rock. ☐
- I know that copper can be extracted by heating its ore with carbon, but purified by electrolysis. ☐
- I know that alloys are mixtures of metals such as copper and zinc, making brass. ☐
- I know that aluminium doesn't corrode when wet as it has a protective layer of aluminium oxide. ☐
- I know that iron rusts in air and water to make hydrated iron (III) oxide. ☐
- I know that toxic carbon monoxide comes from incomplete combustion of petrol or diesel in cars. ☐
- I know that catalytic converters can remove CO by conversion: $2CO + 2NO \rightarrow N_2 + CO_2$. ☐
- I know that a temperature increase makes particles move faster, so increasing the rate of reaction. ☐
- I know that catalyst is a substance which changes the rate of reaction but is unchanged at the end. ☐

P2 Living for the future

- I can explain how energy from the Sun can be used for heating and producing electricity. ☐
- I can describe how a p-n junction produces electricity. ☐
- I can describe how generators produce electricity. ☐
- I can explain why electricity is distributed via the National Grid at high voltages. ☐
- I can explain the advantages and disadvantages of the fuels used in power stations. ☐
- I can calculate power and the cost of using an electrical appliance for a certain time. ☐
- I know why there is background radiation. ☐
- I can list some uses of alpha, beta and gamma sources and relate their use to their properties. ☐
- I can explain how the Earth's magnetic field is produced and its effect on cosmic rays. ☐
- I can describe how the Moon was formed. ☐
- I can explain why the planets stay in the orbits they do around the Sun. ☐
- I can explain how and why we are exploring space through manned and unmanned spacecraft. ☐
- I can describe asteroids and comets and know the importance of constantly checking NEOs. ☐
- I can describe the Big Bang theory and why scientists believe the Universe is still expanding. ☐

Checklist – Additional Science Unit 1

B3 Living and growing

- I can interpret data on DNA fingerprinting for identification.
- I can describe DNA replication.
- I know that food and oxygen diffuse across the placenta.
- I can describe diffusion as the net movement of particles from a region of high to low concentration.
- I know that arteries transport blood away from the heart.
- I know that a patient can reject a heart transplant.
- I know that at fertilisation haploid gametes join to form a diploid zygote.
- I know that body cells are made by mitosis and gametes are made by meiosis.
- I can identify the main stages of human growth.
- I know that shoots are positively phototropic and negatively geotropic; roots are the opposite.
- I can describe the stages in selective breeding.
- I know that genetic engineering is used to make insulin.
- I can describe some advantages and disadvantages of cloned plants.
- I know that cloned animals could be used to produce organs for transplants.

C3 The periodic table

- I know that the nucleus is made up of protons and neutrons, with each having a relative mass of 1.
- I know that electrons surround the nucleus and occupy shells in order. They have almost no mass, 0.
- I know that positive ions are formed by the loss of electrons from the outer shell.
- I know that negative ions are formed by the gain of electrons into the outer shell.
- I know that non-metals combine by sharing electrons, which is called covalent bonding.
- I can draw the 'dot and cross' diagrams of simple molecules such as H_2, Cl_2, CO_2 and H_2O.
- I know that lithium, sodium and potassium react vigorously with water and give off hydrogen.
- I know that group 1 metals have one electron in their outer shell, which is why they're similar.
- I know that chlorine is a green gas, bromine is an orange liquid and iodine is a grey solid.
- I know that halogens gain one electron to form a stable outer shell. This is called reduction.
- I know that in the electrolysis of dilute sulphuric acid, H_2 is made at the cathode and O_2 at the anode.
- I know that when aluminium oxide is electrolysed, Al is formed at the cathode and O_2 at the anode.
- I know that compounds of copper are blue, iron (II) are light green and iron (III) are orange/brown.
- I know that metals conduct electricity as the 'sea' of electrons move through the positive metal ions.

P3 Forces for transport

- I can state and use the formula: speed = distance ÷ time, including a change of subject.
- I can draw and interpret distance-time graphs and speed-time graphs.
- I can use the formula: acceleration = change in speed ÷ time taken, including a change of subject.
- I can state and use the formula: force = mass x acceleration, including a change of subject.
- I can describe the factors that may affect thinking, braking and stopping distances.
- I can state and use the formula: work done = force x distance, including a change of subject.
- I can state and use the formula: power = work done ÷ time, including a change of subject.
- I can use the formula: KE = $\frac{1}{2}mv^2$ and apply it to braking distances and other examples.
- I can interpret data about fuel consumption and explain the factors that affect these figures.
- I can describe and evaluate the effectiveness of various car safety features.
- I can describe how and use forces to explain why falling objects may reach a terminal speed.
- I can recognise that acceleration in free fall (g) is constant.
- I can interpret a gravity ride (roller coaster) in terms of PE, KE and energy transfer.
- I can state and use the formula: weight = mass x g, including a change of subject.
- I can state and use the formula: speed = distance ÷ time, including a change of subject.

Checklist – Additional Science Unit 2

B4 It's a green world

- I can label a diagram showing the parts of the leaf. ☐
- I can explain how a leaf is adapted for photosynthesis. ☐
- I know that osmosis is the movement of water molecules across a partially-permeable membrane. ☐
- I can describe the structure of xylem and phloem. ☐
- I can explain how transpiration rate can be increased. ☐
- I know that active transport needs energy from respiration. ☐
- I can identify mineral deficiencies in plants. ☐
- I can construct pyramids of numbers and biomass. ☐
- I can explain the reasons for developing biofuels. ☐
- I can describe the difference between intensive farming and organic farming. ☐
- I know that a saprophyte decays organisms by releasing enzymes. ☐
- I can explain why the different preservation methods stop food decay. ☐
- I can describe the carbon cycle. ☐
- I can describe the nitrogen cycle. ☐

C4 Chemical economics

- I know that neutralisation is a reaction where: acid + base → salt + water. ☐
- I can construct balanced symbol equations, such as: $2KOH + H_2SO_4 \rightarrow K_2SO_4 + H_2O$ ☐
- I can work out the relative formula mass of a substance from its formula, such as $Ca(OH)_2$. ☐
- I can work out the percentage yield using the formula: % yield = actual yield × 100 ÷ predicted yield. ☐
- I know that fertilisers provide extra essential elements but excess can cause eutrophication. ☐
- I know that ammonia is made by the Haber process where N_2 and H_2 are put over an iron catalyst. ☐
- I know that the higher the pressure, the higher the percentage yield of ammonia. ☐
- I know that a catalyst will increase the rate of reaction but will not change the percentage yield. ☐
- I know that a detergent has a hydrophilic head and a hydrophobic tail. ☐
- I know that dry cleaning is a process used to clean clothes using a solvent that isn't water. ☐
- I know that a continuous process makes chemicals all the time but a batch process doesn't. ☐
- I can recognise the three allotropes of carbon: diamond, graphite and buckminster fullerene. ☐
- I can explain that graphite is slippery and is used as electrodes as it conducts electricity. ☐
- I know that water purification includes filtration, sedimentation and chlorination. ☐

P4 Radiation for life

- I can describe static electricity in terms of the movement of electrons. ☐
- I can explain how static electricity can be dangerous and how the risk of shock can be reduced. ☐
- I can describe and explain some uses of static electricity. ☐
- I can explain the behaviour of simple circuits and how resistors are used in circuits. ☐
- I can state and use the formula: resistance = voltage ÷ current, including a change of subject ☐
- I can explain the function of live, neutral and earth wires, circuit breakers and double insulation. ☐
- I can explain how a wire fuse protects an appliance. ☐
- I can describe the motion of particles in longitudinal and transverse waves. ☐
- I can explain how ultrasound is used in body scans and to break down kidney stones. ☐
- I can explain how nuclear radiation is used in hospitals to treat cancer and as a tracer. ☐
- I can describe radioactive decay and explain and use the concept of half-life. ☐
- I can describe alpha and beta decay and construct nuclear equations to explain it. ☐
- I can explain the origins of background radiation and some non-medical uses of radioisotopes. ☐
- I can describe in detail how domestic electricity is generated in a nuclear power station. ☐

Answers

B1 Understanding ourselves

Page 4
1. glucose → lactic acid + energy.
2. They need to repay the oxygen debt.
3. Breathing and heart rate.
4. Age, diet and amount of exercise.

Page 5
1. $80 \times 0.75 = 60$ g
2. They contain images of thin people that influence a person's view of themselves.
3. They break down large food molecules into small soluble molecules.
4. Emulsification increases the surface area. Digestion breaks down the molecules.

Page 6
1. They carry pathogens from one host to another.
2. Malignant tumours are cancer cells, which divide out of control and spread around the body. Benign cells are slow to divide and harmless.
3. They lock onto antigens causing the bacteria to stick together.
4. In a blind trial, volunteers do not know which treatment they are receiving. In a double blind trial, the doctors also do not know which treatment is used.

Page 7
1. The iris.
2. The ciliary muscles become less flexible.
3. Stimulus, receptor, sensory neurone, CNS, motor neurone, effector, response.
4. Transmitter is only made on one side of the synapse.

Page 8
1. LSD.
2. It stimulates the acetylcholine receptors, allowing more impulses to pass.
3. Cigarette smoke stops the cilia from moving. Dust and particulates collect and irritate the cells. Smokers cough to move this mess upwards so it can be swallowed.
4. They block the exchange of gases and reduce the amount of oxygen available to the rest of the body.

Page 9
1. It is when body temperature falls too low.
2. It is widening of the blood vessels.
3. Hair growth in the pubic area or under the arms.
4. Oestrogen causes the repair of the uterus wall and progesterone maintains it. Together they control ovulation.

Page 10
1. Four.
2. 35.
3. T.
4. DNA codes for an enzyme, and the enzyme controls the reaction that makes the pigment to colour the eye.

Page 11
1. They are genetically identical.
2. One from: radiation, chemicals, they occur spontaneously.
3. Blue.
4.

C1 Carbon chemistry

Page 13
1. They change shape.
2. Irreversibly change shape.
3. Decomposes means that it breaks down. The three products are sodium carbonate, carbon dioxide and water.
4. $2NaHCO_3 \rightarrow Na_2CO_3 + H_2O + CO_2$

Page 14
1. Ascorbic acid (vitamin C).
2. It makes it more difficult for bacteria or mould to grow.
3. They are made up of two parts, a head and a tail. The tail is a 'fat-loving' part and the head is a 'water-loving' part.
4. The hydrophobic tail is attracted into the oil but the head is not. The hydrophilic head is attracted to water and 'pulls' the oil on the tail into the water.

Page 15
1. Acid and alcohol.
2. One person may object to cosmetics being tested on animals, as the animals may be harmed and they have no control over what happens to them. The other may say that they feel safer if the cosmetics have been tested on animals.
3. A substance that evaporates easily.
4. The force of attraction between a varnish molecule and a solvent molecule is stronger than between two varnish molecules. Water doesn't dissolve nail varnish because the force of attraction between two water molecules is stronger than that between a water molecule and a molecule of nail varnish.

Page 16
1. At the top.
2. Bitumen and heavy oil, because there are stronger forces of attraction between longer chains.
3. C_7H_{16}.
4. By cracking large hydrocarbons to make smaller more useful molecules.

Page 17
1. High pressure and a catalyst.
2.
3. Alkanes do not have a double C=C bond, alkenes do.
4. C_nH_{2n}.

Page 18
1. It makes people sweat as it is not breathable.
2. A Gore-Tex® pore is 700 times larger than a water vapour molecule.
3. Toxic gases.
4. They have weak intermolecular forces of attraction between the polymer molecules, so the chains can slide over one another and can be stretched easily.

Page 19
1. Carbon dioxide is a greenhouse gas and contributes to climate change. It is a global problem that cannot be solved by one country and so needs careful agreement.
2. Limewater.
3. Less soot is made, more heat is released and toxic carbon monoxide gas is not produced.
4. $C_5H_{12} + 8O_2 \rightarrow 5CO_2 + 6H_2O$

Page 20
1. When energy is transferred to the surroundings in a chemical reaction (energy is released).
2. A reaction where there is more energy given out when the products are formed than the energy taken in to break the bonds of the reactants.
3. A yellow flame.
4. Energy transferred = $100 \times 4.2 \times 30 = 12600$
Energy per gram = $\frac{12600}{2} = 6300$ J/g

P1 Energy for the home

Page 22
1. Energy is transferred from your hand to the ice.
2. The roof is cool.
3. The specific heat capacity of the syrup is greater than the specific heat capacity of the sponge.
4. 113 000 J.

Page 23
1. 60 years.   2 £120.
3. £97.50.
4. **a** Radiation **b** convection **c** conduction.

Page 24
1. A vacuum does not conduct heat.
2. Less energy is transferred through the ceiling into the roof.
3. Trapped air in the foam does not allow convection into the roof space, and it reduces conduction.
4. Particles vibrate more when heated. Kinetic energy is transferred from one particle to another (conduction) across the brick.

Page 25
1. For safety, as microwaves could warm the water in the human body.
2. They are not in line of sight; in valleys, behind hills or tall buildings.
3. Food is heated by conduction and convection.
4. Longer wavelengths (lower frequencies) transfer less energy – mobile phones use longer wavelength microwaves than those used by microwave ovens.

Page 26
1.

2. Interference is not evident.
3.

4. It reduces the need for using X-rays or surgery.

Page 27
1. There is less refraction of other waves that could interfere.
2.

3. Radio waves are reflected from the ionosphere back down to Earth, and reflected back into the atmosphere for further reflection by the ionosphere by water but not by land.
4. There is diffraction at the edges of the aerial.

Page 28
1. 300 m/s.
2. $1 \rightarrow 10^9$ Hz (1000 MHz or 1 GHz).
3. It uses a series of dots and dashes to represent letters of the alphabet.
4. **a**

b

Page 29
1 The focus is the source of an earthquake. The epicentre is the point on the Earth's surface above the focus.
2 The refraction of P waves gives us information about the size of core. S waves are not detected on the opposite side of Earth, which shows that the core contains liquid.
3 150 minutes (2.5 hours).
4 The ozone layer will be destroyed leading to increased levels of ultraviolet radiation and more cases of skin cancer.

B2 Understanding our environment
Page 31
1 It suffocates.
2 **a** Herbicide **b** pesticide.
3 A group of animals or plants of the same species.
4 Use numbers as coordinates or throw it over your shoulder.

Page 32
1 To trap light or for photosynthesis or make food.
2 They have no chloroplasts or they can't make their own food.
3 So they can survive in water.
4 Felidae.

Page 33
1 Any three from: respiration (energy), storage, make proteins, make cellulose.
2 $6CO_2 + 6H_2O \xrightarrow[\text{(chlorophyll)}]{\text{(light energy)}} C_6H_{12}O_6 + 6O_2$.
3 Keeping the plants warm, providing the plants with extra carbon dioxide, increasing the amount of light for the plant.
4 It is produced from respiration but not needed in photosynthesis.

Page 34
1 So the species survives.
2 There is increased competition from the grey squirrel.
3 A relationship where both organisms benefit.
4 There is more food, so the owls can raise more young.

Page 35
1 To stop sand getting in their eyes.
2 They are large to spread load on snow, and have fur on the sole for grip and insulation.
3 To reduce water loss.
4 To attract insects.

Page 36
1 One from: some body parts decay quickly before they can be fossilised; fossilisation is rare, most living things will completely decay; there may still be fossils we have not found.
2 Organisms were created, they did not evolve.
3 In their genes.
4 Within any species there is variation. There will be competition for limited resources such as food. Only those best adapted will survive, called survival of the fittest. Successful adaptation are passed to the next generation in genes. Over time, the changes may result in a new species. The less well adapted species may become extinct.

Page 37
1 Sulphur dioxide.
2 The USA uses more fossil fuels (oil, gas coal).
3 Any two from: bloodworm, water louse, sludge worm, rat-tailed maggot.
4 It cannot tolerate the low oxygen levels found in polluted water.

Page 38
1 They may be unable to adapt to the change fast enough.
2 For example, local food or dances could be preserved as entertainment for tourists.
3 The whales lose their freedom.
4 For example, prevent the import of wood from unsustainable forests or provide money for local communities to find alternative incomes to deforestation.

C2 Rocks and metals
Page 40
1 A **colloid** is formed when small solid particles are dispersed through the whole of a liquid, but are not dissolved in it.
2 First the solvent evaporates, then the binding medium is oxidised.
3 Emulsion, as the evaporating solvent is water not another type of solvent.

Page 41
1 Limestone and clay.
2 Under heat and pressure.
3 Calcium oxide and carbon dioxide.
4 $CaCO_3 \rightarrow CaO + CO_2$.

Page 42
1 They are less dense.
2 When two plates collide, the more dense oceanic plate sinks under the less dense continental plate. The rocks partially re-melt.
3 If the magma cools slowly, large crystals are made. Rapid cooling produces smaller crystals.
4 Silica-rich magma is less runny than iron-rich magma and produces volcanoes that may erupt explosively shooting out hot ash and pumice.

Page 43
1 The cathode.
2 Copper sulphate solution.
3 Copper and zinc.
4 They change shape.

Page 44
1 The oxide of aluminium becomes a protective layer.
2 Hydrated iron(III) oxide.
3 Aluminium is less dense and so the car is lighter.
4 Aluminium will corrode less or produce a lighter car than steel. Steel will be stronger and cost less to produce.

Page 45
1 Photosynthesis.
2 The process of photosynthesis is reduced so that carbon dioxide is not converted to oxygen.
3 Changes carbon dioxide to carbon monoxide.
4 $2CO + 2NO \rightarrow N_2 + 2CO_2$.

Page 46
1 For a reaction to take place, particles must collide often enough, with sufficient energy. If the particles move faster they will collide more successfully. If they are more crowded they will collide more often. In both cases the reaction will be faster.
2 As the concentration increases, the number of successful collisions per second increases and so the rate of reaction increases.
3 At a higher temperature the particles have more energy, so they collide more successfully, so the reaction is quicker. However, if the same mass of magnesium is used it will produce the same volume of hydrogen each time.
4 Extrapolation means extending a graph to read an expected reading that has not been measured.

Page 47
1 Any two from: sulphur, flour, custard powder, wood dust.
2 Half the mass of zinc will produce only half the volume of gas if the acid remains in excess both times.
3 There will be more collisions each second, which means the rate of reaction increases.
4 It helps reacting particles collide with the correct orientation and allows collisions between particles with less kinetic energy than normal to be successful.

P2 Living for the future
Page 49
1 No mains electrical supply needed in remote areas.
2 n-type silican has an excess of free electrons, p-type silicon has an absence of free electrons.
3 Sun over the equator – the equator is north of Australia but south of England.
4 Infrared.

Page 50
1 Steam is pressurised.
2 There is less energy loss and lower distribution costs.
3 High voltage means a lower current. There's less heating effect and less energy lost to the environment.
4 0.3 or 30%.

Page 51
1 5p
2 One from: washing machine, dishwasher, night storage heater.
3 Radiation causes ionisation, which changes the structure of atoms. DNA in the cell can change, so the cell behaves differently and divides in uncontrolled way. This causes cancer.
4 It conserves fossil fuels for other uses, and as no carbon dioxide is produced there is no contribution to climate change.

Page 52
1 75%.
2 The nucleus contains positive protons.
3 Alpha radiation would not penetrate. Gamma radiation would give little change in the count rate when thickness changes slightly.
4 There is a risk of damage to the store or the container, which would result in a radiation leakage.

Page 53
1 They fluoresce.
2 From a collision between two planets.
3 The Moon has the same oxygen content as the Earth.
4 Cosmic rays interacting with the Earth's magnetic field near the poles.

Page 54
1 Mercury, Venus, Earth, Mars, Jupiter, Saturn, Uranus, Neptune, (Pluto).
2 Planet moves in a straight line at a tangent to the circular orbit.
3 To keep the body at a suitable temperature, maintain a suitable pressure, provide oxygen for breathing.
4 **a** 144 000 000 km **b** $\times 10^{13}$ km.

Page 55
1 Large rocks or small planets orbiting the Sun.

2 Dust from a collision with an asteroid blocked out the Sun. Herbivorous dinosaurs could not feed since plants could not photosynthesise. So the carnivorous dinosaurs had no food.
3 In case their orbit passes close to the orbit of the Earth.
4 A rocket filled with explosive could be sent to hit the NEO and alter its course.

Page 56
1 The galaxies furthest away.
2 Red shift gives indication of the speed of a star. Use speed and distance to calculate time.
3 The Sun was formed from remnants of a supernova, which was the result of a former star exploding.
4 They use hydrogen at a slower rate than large stars.

B3 Living and growing
Page 58
1 Amino acids are joined together.
2 Three.
3 An enzyme is a protein that acts as a biological catalyst.
4 It denatures the enzyme, changing the shape of the active site.

Page 59
1 The movement of molecules from a high to a low concentration.
2 They increase the surface area, which speeds up diffusion.
3 It carries a signal from one neurone to the next.
4 It diffuses through the stomata.

Page 60
1 Haemoglobin.
2 Pulmonary artery.
3 To withstand high pressure.
4 Blood going to the body can be pumped at a much higher pressure than blood going to the lungs.

Page 61
1 If the cell is too large it could not absorb enough food and oxygen through the surface of its membrane to stay alive.
2 23 pairs or 46.
3 Meiosis.
4 Both make new cells, at one point single strands move to opposite poles.

Page 62
1 Any two from: cell wall, chloroplast, large vacuole.
2 They are easier to grow.
3 They are bigger so they need time to develop enough to survive outside the uterus.
4 Digestive system problems.

Page 63
1 It makes roots grow.
2 The root.
3 It causes elongation of the cells.
4 It causes elongation of the cells on the dark side.

Page 64
1 Choose the characteristic. Cross-breed. Select the best offspring. Repeat the selection and breeding process for a number of generations.
2 There is a reduction in variation; more chance of harmful recessive genes being expressed.
3 There may be harmful effects on humans who eat them.
4 Select the characteristic. Identify and isolate the gene. Insert the gene into chromosome of a different organism. Replicate (copy) the gene in the organism and produce the protein.

Page 65
1 Embryo transplantation.
2 Nuclear transfer.
3 One from: the plants are all genetically identical, if the environment changes or a new disease breaks out it is unlikely that any of the plants will survive; cloning plants over many years has resulted in little genetic variation.
4 Plants with the desired characteristics are chosen. A large number of small pieces of tissue are taken from the parent plant. They are put into sterile test tubes that contain growth medium. The tissue pieces are left in suitable conditions to grow into plants.

C3 The periodic table
Page 67
1 1.
2 Chlorine-37.
3 11.
4 Magnesium.

Page 68
1 Electrons are gained.
2 A metal atom needs to lose electrons. The electrons transfer from the metal atom to a non-metal atom. A non-metal atom needs to gain electrons. The electrons transfer to the non-metal atom from the metal atom.
3

LiF
4 Because ions can move in the molten liquid.

Page 69
1

2 H $\overset{\times \times}{\underset{\times \times}{\times \overset{}{O} \times}}$ K
H$_2$O
3 The molecules are easy to separate.
4 Third.

Page 70
1 It is less dense than water.
2 Group 1 metals all have 1 electron in their outer shell.
3 Rubidium hydroxide.
4 The fourth shell is further away from the attractive 'pulling force' of the nucleus, so the electron from potassium is more easily lost than the electron from sodium. Potassium is therefore more reactive than sodium.

Page 71
1 The process of electron gain.
2 Potassium + bromine → potassium bromide.
3 Iodine is less reactive than bromine so does not displace it.
4 $2Li + Br_2 → 2LiBr$.

Page 72
1 The formula of the compound breaking up is H_2O.
2 The H^+ ions migrate towards the cathode where they accept an electron. Two ions bond together to form a molecule of hydrogen gas, H_2.
3 They are gradually worn away by oxidation.
4 To lower the temperature of the melting point of aluminium oxide to save energy costs.

Page 73
1 Orange/brown.
2 In sodium hydroxide solution, Fe^{2+} ions form a grey/green solid and Fe^{3+} ions form an orange gelatinous solid.
3 copper carbonate → copper oxide + carbon dioxide.
4 $Cu^{2+} + 2OH^- → Cu(OH)_2$.

Page 74
1 It has a high thermal conductivity and is resistant to attack by metals or acids.
2 A metallic bond has a strong electrostatic force of attraction between close-packed positive metal ions and a 'sea' of delocalised electrons.
3 It levitates.
4 They only work at very low temperatures.

P3 Forces for transport
Page 76
1 30 m/s.
2 7 hours.
3 a It is constant
  b it is decreasing.
4 Its speed is 25 m/s.

Page 77
1 The distance travelled.
2 From the gradient of the graph.
3 2.5 m/s^2.
4 24 m/s.

Page 78
1 6000 N.

2 The forces are the same size, are in opposite directions and act on different objects (the Earth and the Moon).
3 14 m.
4 The brakes provide a backwards force to slow the car down. This backwards force is provided by the friction of the brakes against the wheels. Worn brakes provide a smaller friction force so increase braking distance.

Page 79
1 1600 J.
2 5000 N; heat in brakes (mainly).
3 400 W.
4 180 000 J; e.g. motor not 100% efficient, energy wasted as heat due to friction, etc.

Page 80
1 A larger mass means the car must gain more kinetic energy, so it uses more fuel.
2 Any three from: reduce speed, avoid rapid acceleration, avoid rapid deceleration, reduce gear changes, reduce use of brakes, don't use a roof rack.
3 a 60 000 J b 12 m.
4 Any from: short range, slow speed, batteries take time to recharge, need a recharging facility, batteries take up a lot of space, batteries are heavy.

Page 81
1 Any two from: thermal energy, elastic energy, sound energy.
2 This is less tiring on long motorway journeys. It helps a car keep to a steady speed so the driver does not have to use the pedals.
3 25 m.
4 Increasing collision time reduces the acceleration, and therefore the forces acting on the people in the car.

Page 82
1 Weight and air resistance/drag. a Weight constant, drag = 0 b weight = drag c drag greater than weight.
2 As the parachutist falls, air molecules are displaced causing an upward force – air resistance/drag. The faster he/she falls the greater this drag force. This reduces the net force acting, F (= weight – air resistance). $F = ma$ so if F is less, a (acceleration) is less.
3 More air molecules are displaced each second.

Page 83
1 Gravitational PE → KE → gravitational PE, etc.
2 4000 J.

3 It has very little KE at the top (mainly gravitational PE) but a lot of KE at the bottom.
4 **a** 60 kg **b** 1.67 N/kg.

B4 It's a green world
Page 85
1 Palisade cells.
2 They allow gases to diffuse easily and reach all the cells.
3 So it has a large surface area to absorb light.
4 To allow gas exchange.
Page 86
1 Osmosis.
2 Water leaves the cells, which become plasmolysed.
3 Less sugar is made, water leaves the guard cell, the guard cell becomes plasmolysed, the stoma closes.
4 To reduce water loss.
Page 87
1 Xylem vessels.
2 They contain lignin.
3 Any two from: increased light intensity, increased temperature, increased air movement (wind), a fall in humidity (amount of water in the atmosphere).
4 When light intensity increases the stomata open. This allows more water to escape.
Page 88
1 Potassium and magnesium.
2 Phosphorus.
3 Magnesium.
4 Active transport requires energy and is against the concentration gradient; it also uses carrier molecules.
Page 89
1 A pyramid of numbers shows the number of organisms, a pyramid of biomass shows the mass of the organisms.
2 $\frac{40}{2000}$ = 0.02 or 20%.
3 Sugar cane is fermented using yeast to make alcohol. The alcohol is mixed with petrol to make Gasohol, a fuel for cars.
4 The plants used can be replaced by growing more.
Page 90
1 Suggestions such as: enclosed in very small spaces, given harmful drugs, etc.
2 The mineral supply is controlled and unused minerals are recycled, reducing costs. There is better control of external conditions and disease.
3 One from: grow nitrogen-fixing plants, use manure, use compost.
4 One from: high costs, small yield.
Page 91
1 Earthworms, maggots and woodlice.
2 The temperature at which enzymes work best.

3 An organism that feeds off dead and decaying material.
4 It removes the moisture that bacteria need for growth.
Page 92
1 Photosynthesis.
2 Limestone reacts with acid rain to release carbon dioxide.
3 A decomposer.
4 One from: nitrogen-fixing bacteria convert it into ammonia or nitrates; lightning combines nitrogen and oxygen to form oxides, the oxides dissolve in rain and form nitrates in the soil.

C4 Chemical economics
Page 94
1 Zinc oxide, zinc hydroxide or zinc carbonate.
2 Magnesium sulphate.
3 $ZnO + 2HCl \rightarrow ZnCl_2 + 2H_2O$.
4 Hydroxide ions, OH^-.
Page 95
1 $40 + 2(14 + (3 \sqrt{16})) = 40 + 2(14 + 48) = 40 + 2(62) = 40 + 124 = 164$.
2 When chemicals react, the atoms of the reactants swap places to make new compounds, which are the products. They do not disappear.
3. 75%.
4. 22 g CO_2.
Page 96
1 149.
2 The percentage that is nitrogen is $\frac{42}{149} \times 100 = 28\%$.
3 Potassium hydroxide and phosphoric acid.
4 Titrate the potassium hydroxide with phosphoric acid, using an indicator. Repeat the titration until three consistent results are obtained. Use the titration result to add the correct amounts of acid and alkali together, without the indicator. Evaporate most of the water using a hot water bath. Leave the remaining solution to crystallise, then filter off the crystals.
Page 97
1 About 450°C.
2 The yield is lower than at a lower temperature but the rate of production is higher.
3 The yield would be too low.
4 Very high pressures are difficult and costly to maintain. A lower, optimum pressure is chosen to give a reasonable yield at a reasonable cost.
Page 98
1 An organic acid and an alkali.
2 The oil.

3 Fabrics may be damaged by washing in water and grease stains do not dissolve in water but do in dry-cleaning solvent.
4 Water to water.
Page 99
1 A batch process is where the whole process takes a limited time then stops and can be changed if necessary.
2 The plant can be used to full capacity at all times.
3 Any two from: there are legal requirements, investment costs of its research and development, raw materials, expensive extraction from plants, it is labour intensive.
4 Many years.
Page 100
1 Conducts electricity and has a high melting point.
2 Any two from: semiconductors in electrical circuits, industrial catalysts, reinforce graphite for tennis rackets.
3 Graphite has delocalised electrons that can move through the structure. Diamond does not.
4 To trap or transport molecules.
Page 101
1 Sand is used to filter out fine particles that do not sediment out.
2 The water needs to be heated up to boiling point and then cooled.
3 Precipitation.
4 $2AgNO_3(aq) + MgBr_2(aq) \rightarrow 2AgBr(s) + Mg(NO_3)_2(aq)$.

P4 Radiation for life
Page 103
1 **a** They repel **b** they attract.
2 Electrons move from the acetate rod to the duster.
3 They could cause a spark which may lead to an explosion as fuel is highly inflammable.
4 The sheets contain oil so they reduce friction (so fewer electrons are transferred); they are made from conducting material so charge cannot build up.
Page 104
1 To ensure good electrical contact.
2 5 ms (0.005 s).
3 Electrons flow to or from earth to keep the frame uncharged giving a more even coverage; there is less waste; shadows painted.
4 To charge the soot particles when they come near the wires.
Page 105
1 8 Ω.
2 0.4 A.

3 Wire in a fuse melts if the current becomes too large, breaking the circuit and preventing overheating.
4 It is connected to the case of an appliance so that if the case becomes 'live', a large current flows in the earth and live wires and the fuse 'blows'.
Page 106
1 Body fat.
2 Transverse.
3 10 cm.
4 0.24 m (24 cm).
Page 107
1 Using radiation to treat diseases such as cancer.
2 Similarities, two from: both electromagnetic radiation, both have very small wavelength, both very penetrative, both damage or kill living cells. Differences, two from: obtained in different ways, X-rays easy to control but gamma are not, gamma rays can be used as a tracer but X-rays cannot, gamma rays can sterilise equipment but X-rays cannot.
3 Iodine is taken up by the thyroid gland and it emits only gamma radiation.
4 So that the tumour receives the full dose of radiation but healthy tissue receives less radiation.
Page 108
1 Radioactive decay is a random process.
2 4 hours.
3 $^{215}_{84}Po$.
Page 109
1 Iron was never living so does not contain carbon-14.
2 Only gamma radiation will penetrate through pipes, ground, etc.
3 Less uranium has decayed to lead (i.e. there is more uranium and less lead than an old rock).
4 Very little carbon-14 would have decayed so there would be very little change in activity.
Page 110
1 Uranium containing a greater proportion of the uranium-235 isotope than occurs naturally.
2 Extra neutrons are released which split more uranium nuclei producing more neutrons and so on.
3 It emits harmful ionising radiation for a long time.
4 They absorb some of the neutrons to control the number of fissions occurring.

Glossary and index

buckminster fullerene	A very stable ball of 60 carbon atoms joined by covalent bonds; the whole structure looks like a geodesic dome.	100, 102
cancer	A dangerous illness caused by radiation, smoking and some types of chemical	6, 29, 30, 51, 107
capillary	The narrowest type of blood vessel, so narrow that red blood cells can only pass through in single file.	9, 60, 66
carbohydrate	Carbohydrates are chemicals found in all living things. They contain the elements carbon, hydrogen and oxygen.	5, 13, 21
carbon	A very important element, carbon is present in all living things and forms a huge range of compounds with other elements.	17, 44, 92, 93, 100, 102
carbon dioxide	A gas containing only carbon and oxygen. Its chemical formula is CO_2.	13, 19, 21, 33, 37, 45, 48, 51, 59, 66, 69, 73, 75, 79, 85, 92, 93
carbon monoxide	A poisonous gas containing only carbon and oxygen.	19, 45
carbon-14	A radioactive isotope of carbon used in carbon-dating calculations.	109
carbonate	Compounds containing the carbonate group of atoms; the carbonate group formula is CO_3.	13
carnivore	An animal that eats other animals.	89
carrier	An organism that is carrying a pathogen of some sort without necessarily suffering from the disease.	88
carrier wave	An electromagnetic wave that can carry speech, music, images, or other signals.	26
catalyst	A chemical that speeds up a reaction but is not changed or used up by the reaction.	14, 17, 45, 47, 58, 66, 73, 75, 97
catalytic converter	Boxes fitted to vehicle exhausts which reduce the level of nitrogen oxides and unburnt hydrocarbons in the exhaust fumes.	45
cathode	The negative electrode in a circuit or a battery.	43, 72, 75
cell differentiation	The change of an unspecialised cell into a particular type of cell.	61
cell division	The multiplication of cells.	61, 62, 66
cell membrane	The thin layer around the outside of a cell.	62, 88
cell wall	A thick wall consisting of cellulose and starch found around plant cells.	62, 66, 86
cellulose	An insoluble carbohydrate molecule formed of spirals of glucose joined together. Cell walls in plants are made of cellulose.	33, 62
central nervous system	The brain and spinal cord.	7
chain reaction	A reaction where the products cause the reaction to go further or faster, e.g. in nuclear fission.	110
characteristic	Characteristics are things that are true about something, e.g. it is a characteristic of mammals that they have fur.	11, 19, 64, 65
charge	A property of matter, charge exists in two forms (negative and positive) which attract each other.	52, 67, 103, 104, 105
chemical change	A change that occurs when a number of substances react together to produce new substances.	21
chlorination	Adding chlorine to a molecule or substance.	101
chlorophyll	A green chemical found in photosynthetic plants that allows them to use energy from sunlight to make sugar from carbon dioxide and water.	85, 88, 93
cholesterol	A chemical found particularly in animal fats and foods made from them; the body uses cholesterol to make nerve tissue and some hormones but high levels of cholesterol may make heart attacks more likely.	60
chromosome	A thread-like body made of DNA and protein found in the nucleus. It only becomes visible during cell division.	10, 11, 12, 61, 66
cilia	Tiny hairs found on the surface of cells and some microorganisms.	8
circuit	The complete path around an electrical circuit that electricity can flow along.	105, 111
circuit-breaker	A device which breaks a circuit to switch off the power when a danger is detected, it behaves similarly to a fuse.	105
cirrhosis	A disease of the liver often caused by excessive alcohol intake.	8
classification	Arranging things into sets according to the features that they have in common.	32, 39
clone	Two living thing are clones if they have exactly the same genes.	65
collision frequency	The number of collisions in a particular area in a particular time.	46, 47, 48
colloid	A mixture in which small particles of one substance are suspended in another.	40, 48
combustion	The reaction between a fuel and oxygen to form carbon dioxide and water, and release energy as light and heat.	19, 20, 21, 48
comet	A small body that moves through space and reflects light.	54, 55
compost	Organic material, typically kitchen and garden waste, being rotted down to make fertiliser for the soil.	90, 91
compound	Compounds are groups of atoms bound together, in fixed proportions, by chemical bonds. Compounds have different chemical and physical properties to the elements that they contain.	43, 73
concentration	The amount of material in a given volume.	46, 59
concentration gradient	The difference in concentration of a substance between two areas divided by the distance between them.	88
conduction	Movement of energy as heat (or electricity) through a substance without the substance itself moving.	23, 24, 25, 30, 74, 75
continuous process	A process which can continue indefinitely if new materials are added and wastes removed.	99, 102
control rod	Metal rods which are lowered into nuclear reactors to control the rate of decay and so the temperature of the core.	110

Term	Definition	Pages
convection	Movement of a heated substance which carries energy with it as heat.	23, 24, 25, 30
convection current	Movement upwards of heated gases or liquids to float on top of the cooler, denser layers.	42
coronary artery	The artery that supplies blood to the muscles of the heart.	60
corrode	Metals react with the air to form powder or crystals and weaken the metal, the word rust is used when iron corrodes.	44
cosmic ray	Radiation from space which hits the atmosphere. Some passes through while some is blocked.	53, 109
covalent bond	A link between two atoms where electrons move around both atoms in the pair; covalent bonds tend to be strong bonds and form between non-metals.	17, 69, 75, 100
cracking	The breaking of large organic molecules into smaller ones using heat, pressure and sometimes catalysts.	16
critical angle	The angle below which light is reflected back into a glass block.	26
crop yield	The mass of useful material produced by a crop.	96
cross-breed	To produce an organism by mating individuals from different breeds, varieties, or species.	64
cryolite	A compound of fluorine, aluminium and sodium (Na_3AlF_6) used to extract aluminium from bauxite by electrolysis.	72
crystals	A solid substance with a regular shape made up of flat planes.	42, 96
current	The flow of electrical charge through an electrical circuit.	49, 50, 53, 57, 72, 103, 105, 111
cuticle	The waxy outer layer of the skin or the surface of a leaf.	35, 86
cyclic fluctuation	A variation which rises and falls regularly, for example the numbers of predators and their prey.	34, 39
cytoplasm	The material in the cell, inside the cell membrane but outside the nucleus.	62, 66
decay	To rot.	36, 91, 93, 108
deceleration	Reduction of the velocity of a moving object; it is usually calculated as change in velocity per second and given a minus sign.	77, 82
decompose	To break apart.	73, 92
decomposer	An organism that breaks down dead organic matter.	92, 93
delocalised electron	An electron in a molecule that is not linked directly to an individual atom or covalent bond.	74
denatured	A denatured protein molecule has had its shape altered, e.g. by heat, so that it cannot do its job efficiently.	13, 58
denitrifying bacteria	Bacteria that breakdown complex nitrogenous substances and release nitrogen gas.	92, 93
density	The mass of an object divided by its volume.	53, 70
deoxyribonucleic acid	The molecule that carries the genetic information in animals and higher plants. Often abbreviated to DNA.	10
detritivore	An organism that eats dead material, e.g. an earthworm.	91
detritus	The dead and semi-decayed remains of living things.	91
diabetes	A disease caused by the failure to control blood sugar levels due to the inability of the pancreas to secrete insulin.	5, 6, 9, 64, 65
diastolic pressure	The pressure of the blood when the ventricles of the heart relax. It is generally lower than systolic pressure.	4, 12
differentiation	A process in which new cells develop specialised features in order to carry out a job.	61, 62, 65, 66
diffraction	A change in the directions and intensities of a group of waves after passing by an obstacle or through an opening whose size is approximately the same as the wavelength of the waves.	25, 27
diffusion	The spreading of gases or liquids caused by the random movement of their molecules.	5, 15, 59, 60, 66, 85, 86, 93
digestive system	The gut and the other body parts linked to it that help with digestion.	5
diploid	Cells of organisms which contain chromosomes in pairs.	61, 66
displacement reaction	A reaction where one chemical, usually a metal, is forced out of a compound by another chemical, also usually a metal.	71
displayed formula	A displayed formula shows all the bonds in the molecule as individual lines.	17
dissolve	When a solid mixes in with a liquid so that it cannot be seen.	18, 43, 86, 96
distillation	Scientists use distillation to boil off a liquid from a mixture and then cool the vapours to produce a purer liquid.	16, 21, 101
DNA	The molecule that carries the genetic information in animals and higher plants. DNA is short for Deoxyribonucleic Acid.	10, 11, 12, 51, 58, 64, 66, 88
DNA replication	Copying of a DNA molecule to produce another, identical molecule.	58
dormant	A plant or seed is dormant if it is still alive but fails to grow even though the correct conditions for growth are available.	63
'dot and cross' model	A way to show how electrons are shared in covalent bonds.	68, 69
double circulatory system	A circulatory system where the blood passes through the heart twice – once for the body and once for the lungs.	60
double helix	Two helices wound around each other, DNA is a molecule with a double helix structure.	10
dust precipitator	A device that uses electrostatic charge to make dust in the air fall out of the air.	104
dynamo	A device that converts energy in movement into energy in electricity.	50, 53, 57
earthed	To be connected to the earth – an important safety feature of many electrical appliances.	103
ecological niche	The place or function of a given organism within its ecosystem.	34
ecosystem	The collection of different organisms in an area together with the important non-living factors such as water supply and temperature.	31, 34, 39
efficient	An efficient device transfers most of the input energy into the desired output energy.	89, 90

genetic engineering	A range of technologies that allow scientists to manipulate individual genes.	64, 66
genetically identical	Having exactly the same genes, i.e. being clones.	61
geotropism	Growing towards the centre of the earth.	63, 66
germination	The first stages of growth of a seed into a new plant.	63
gestation	The time between conception and birth in humans, gestation is called pregnancy and lasts for roughly 42 weeks.	62
gland	An organ in the body that produces a secretion.	9
glucose	A type of sugar. Glucose is sometimes called dextrose.	4, 9, 33, 58
gradient	A slope or difference in measurements between two areas, for example there is a concentration gradient between water molecules inside and outside a cell.	46, 47, 76, 77
graphite	A type of carbon often used in pencils as the 'lead'.	75, 100, 102
gravity	The force of attraction between two bodies caused by their mass.	63, 83, 84
greenhouse gas	Gases like carbon dioxide and water vapour that increase the greenhouse effect.	19, 51, 79
Haber process	The industrial process developed by Fritz Haber to make ammonia from nitrogen and hydrogen.	73, 97, 102
haemoglobin	A complex chemical found in red blood cells that can combine with oxygen to help transport it around the body.	60
half-life	Time taken for half of the radioactive atomic nuclei in a sample of an element to break down; the shorter the half-life the more quickly a radioactive chemical decays.	108, 111
halogen	A group of reactive non-metals with only one electron missing from their outer electron shell, e.g. chlorine and iodine.	71, 75
haploid number	The number of chromosomes present in the sperm or egg of a species.	61, 66
herbicide	A chemical that can kill plants.	31, 64, 93
herbivore	An animal that eats plants.	89
heterozygous	An individual who has two different alleles for an inherited character.	11
homeostasis	All living organisms attempt to maintain the conditions in their cells within certain limits. This is known as homeostasis.	9
homologous	A matched pair of chromosomes; humans have 22 homologous pairs of chromosomes and a pair of sex chromosomes.	61
homozygous	An individual who has two alleles that are the same for an inherited feature.	11
hormone	A chemical produced by an endocrine gland which changes the way other parts of the body work; hormones pass around the body in the blood.	9, 63, 66
host	An organism that is carrying another one inside its body.	6, 34
hybrid	An organism made when two different species breed together.	32
hydrocarbon	Hydrocarbon molecules are molecules that contain only carbon and hydrogen atoms.	16, 17
hydrogen	A colourless, odourless gas that burns easily in oxygen to form water; hydrogen is the lightest element.	17, 69, 72, 75, 94, 97, 102
hydrophilic	A molecule or part of a molecule that dissolves easily in water, hydrophilic means 'water loving'.	14, 98
hydrophobic	A molecule or part of a molecule that does not dissolve easily in water, hydrophobic means 'water hating'.	14, 18, 98
hydroponics	Growing plants in mineral solutions without the need for soil.	90, 93
hydroxide	Chemicals containing an 'OH' group; hydroxides are often alkaline.	70, 94
igneous	Rocks formed from solidified molten magma.	41, 42
immune system	The parts of the body that protect against illnesses.	6, 12
inbred	An inbred population has a large number of genes that are the same.	64
indicator	A chemical that changes colour in acid and alkaline solutions; indicators are used to find out the pH of a solution.	96
indicator species	A species that is particularly sensitive to an environmental pollution. The presence or absence of an indicator species is often used to assess the degree of pollution in an environment.	37, 39
infrared	Radiation beyond the red end of the visible spectrum. Infrared radiation is efficient at transferring heat.	25, 30
inherit	To receive something from your parents, usually used to describe characteristics that can be passed down through sperm and eggs.	11, 12
insecticide	A chemical that can kill an insect.	6, 93
insoluble	A substance that will not dissolve.	15, 33
insulation	A substance that slows down the movement of energy.	23, 24, 30, 35, 105
intensive farming	Farming that uses a lot of artificial fertilisers and energy to produce a high yield per farm worker.	90, 93
interference	Waves interfere with each other when two waves of different frequencies occupy the same space. Interference occurs in light and sound and can produce changes in intensity of the waves.	26, 27
intermolecular force	A force between two molecules.	16, 18, 69
invertebrate	An animal without a backbone.	39
ion	Charged particle made when an atom, or group of atoms, gains or loses electrons.	68, 70, 75, 88, 94
ionic equation	An equation showing the movement and behaviour of ions in a reaction.	70, 71
ionisation	The formation of ions (charged particles).	51, 52, 107
isotope	One of two or more atoms having the same atomic number but different mass numbers.	67, 109, 110
joule	A unit of energy(J).	20, 84
kilowatt	1000 watts.	51

Term	Definition	Pages
synthetic	Made by human beings, for example plastics are synthetic compounds which do not occur naturally in nature.	9, 63
systolic pressure	The pressure of the blood when the ventricles of the heart contract. It is generally higher than diastolic pressure.	4, 12
tectonic plate	Sections of the Earth's crust that float on top of the mantle. Plates are hundreds of miles across and move relative to each other by a few inches a year.	42, 48
terminal speed	The speed at which the force of gravity and the force due to air resistance are equal and the object is falling as fast as possible.	82, 83, 84
thermal decomposition	Breaking down a chemical using heat.	41, 73
thermal energy	Energy that can raise the temperature of an object, sometimes called 'heat'.	83
thermochromic	A pigment that changes colour when it gets hotter or colder.	40
thermogram	A picture showing differences in surface temperature of a body.	22
tissue	A group of cells of the same type, so nervous tissue contains only nerve cells.	62, 65, 106, 107
titration	Adding carefully measured amounts of a solution of known concentration to an unknown one to reach an end point which will allow the researcher to calculate the concentration of materials in the unknown solution.	96
tracer	A radioactive, or radiation-emitting, substance used in a nuclear medicine scan or other research where movement of a particular chemical is to be followed.	52, 107, 109
transition element	A metal belonging to the transition group in the periodic table.	73
transpiration	The release of water vapour from a plant through the leaves.	86, 87, 93
transverse	In transverse waves, the vibration is at right angles to the direction in which the wave travels.	29
trophic	The level at which an organism gets its food, primary producers are level one, primary consumers are level two and secondary consumers are level three.	89, 93
turbine	A device that converts movement in a fluid into circular movement, usually to drive a generator.	49, 50, 110
turgid	Plant cells which are full of water with their walls bowed out and pushing against neighbouring cells are turgid.	86
ultrasound	Sounds which have too high a frequency for humans to hear (above 20 khz).	106, 111
ultraviolet	Radiation just beyond the blue end of the spectrum of visible light.	29, 30
universal indicator solution	An indicator that changes colour in solutions of different pH.	94
Universe	Everything, everywhere.	56, 57
unsaturated	An unsaturated solution can dissolve more solute. An unsaturated hydrocarbon can react with more hydrogen because it contains a number of double-carbon bonds.	17
uranium	A radioactive metal used in nuclear power stations and bombs.	109, 110
uterus	The organ in the female where the baby grows during pregnancy, also known as the womb.	9, 62
vacuole	A sac in a cell filled with a watery solution, plant cells tend to have large vacuoles but animal cells have small ones.	62, 66
variable resistor	A resistor whose resistance can change.	105
variation	The existence of a range of individuals of the same group with different characteristics.	11, 61, 64
vascular bundle	A collection of xylem and phloem vessels in a plant, they can be seen in leaves as the veins.	87
vein	In animals: a blood vessel carrying blood towards the heart. In plants: a collection of xylem and phloem vessels clearly seen on the surface of a leaf.	60, 66
velocity	Velocity is the speed an object is moving in a particular direction, usually measured in metres per second (m/s).	77
ventricle	Large muscular chamber in the heart.	60
vertebrate	An animal with a boney backbone or spine.	39
villi	A small projection on the inner surface of the gut to increase the surface area and so speed up absorption.	59
volt	The international unit of electrical potential (V).	105
voltage	The potential difference across a component or circuit.	50, 57, 103, 105
voltmeter	A meter that measures the voltage between two points.	105
watt	A unit of power, 1 watt equals 1 joule of energy being transferred per second (W).	51
wavelength	The distance between two identical points on a wave.	25, 26, 27, 28, 56, 106, 111
weight	The force of gravity acting on a body on Earth; weight is a force and is measured in newtons.	82, 83, 84
work	Work is done when a force moves, the greater the force or the larger the distance the more work is done.	79, 83, 84
X-ray	Electromagnetic radiation used by doctors to look inside a patient's body or to destroy some types of cancer cells.	106, 107, 111
XX chromosomes	The sex chromosomes present in a human female.	11
XY chromosomes	The sex chromosomes present in a human male.	11
xylem	Cells specialised for transporting water through a plant; xylem cells have thick walls, no cytoplasm and are dead, their end walls break down and they form a continuous tube.	87, 93
yeast	A unicellular fungus used extensively in the brewing and baking industries.	89
yield	The ratio of product to starting materials, a high yield means that most of the starting material is converted to useful products.	95, 97, 102
zygote	A cell produced when a male and female gamete join.	61, 66

Collins Revision

GCSE Higher Science

Exam Practice Workbook

FOR OCR GATEWAY B

Fit for life

1 Ben goes running to get fit.

   a After a while Ben's muscles start to hurt.

      i Name the chemical made in Ben's muscles that causes the pain.

      _____ [1 mark]

      ii Explain why the muscles make this chemical.

      _____

      _____

      _____ [3 marks]

   b After the run Ben continues to breathe heavily and his heart rate stays high.
Explain why.

   _____

   _____

   _____ [2 marks]

D–C

B–A*

2 Ben is an athlete and he is very fit. However, Ben still catches a cold.
Explain why being fit may not keep you healthy.

_____

_____ [2 marks]

D–C

3 Amelia has her blood pressure checked by a nurse. The nurse tells Amelia her blood pressure is too high. She fills in a questionnaire for the nurse.

Blood pressure questionnaire

Questions	Notes	Answers Yes	No
1 Do you take regular exercise?	Strong heart muscles will lower blood pressure		✓
2 Do you eat a healthy balanced diet?	Reducing salt intake will lower blood pressure		✓
3 Are you overweight?	Being overweight by 5 kg raises blood pressure by 5 units	✓	
4 Do you regularly drink alcohol?	A high alcohol intake will damage liver and kidneys	✓	
5 Are you under stress?	Relaxation will lower blood pressure	✓	

   a Suggest **two** changes Amelia should make to lower her blood pressure.

     1 _____

     _____

     2 _____

     _____ [2 marks]

   b Describe the possible consequences of high blood pressure for Amelia.

   _____

   _____ [2 marks]

D–C

B–A*

What's for lunch?

1 African children sometimes have a swollen abdomen, a condition caused by a low protein diet.

   a Write down the name of this condition.

   _____ [1 mark]

   b A child has a mass of 40 kg. Calculate his recommended average protein intake (RDA) in grams. Use this formula: RDA in g = 0.75 x body mass in kg
Show your working.

        RDA = _____ g   [2 marks]

2 Simon is overweight.

   a Simon would like to lose weight. He decides to change his diet. Suggest **one other** way Simon could lose weight.

   _____ [1 mark]

   b Simon is a vegetarian. He needs to think carefully about his protein intake. Explain why.

   _____

   _____

   _____ [2 marks]

3 Enzymes are used to digest food.

   a Finish the table to name the enzyme that digests each type of food and the product of the digestion. The first one has been done for you.   [4 marks]

food type	enzyme	product
starch	_carbohydrase_	_glucose_
protein		
fat		

   b Use ideas about **enzymes**, **bile** and **size of droplets** to explain how fats are digested in the body.

   _____

   _____

   _____

   _____ [3 marks]

Keeping healthy

1 Look at the diagram. It shows how mosquitoes spread malaria.

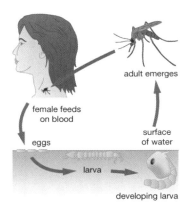

adult emerges

female feeds
on blood

surface
of water

eggs

larva → developing larva

a **i** What name is given to animals, such as the mosquito, that carry pathogens?

_____ [1 mark]

ii Malaria is caused by a parasite called *plasmodium falciparum*.
Explain why it is a parasite.

_____ [1 mark]

b Use the diagram to explain **one** way in which the spread of malaria could be controlled.

_____
_____ [2 marks]

2 a Explain the difference between active and passive immunity.

_____
_____
_____ [3 marks]

b Vaccinations can contain harmless forms of a pathogen. Vaccinations can cause active immunity. Explain how.

_____
_____ [2 marks]

c Antibiotics are drugs used to treat some infections.

i Explain why antibiotics cannot be used to treat a viral infection.

_____ [1 mark]

ii Suggest a reason why doctors are concerned about the over-use of antibiotics.

_____ [1 mark]

iii Doctors need to develop new drugs such as antibiotics. Describe how a blind trial can be used to test a new drug.

_____
_____
_____ [2 marks]

Keeping in touch

1 a Finish the table to show which part of the eye performs which function.
The first one has been done for you. **[3 marks]**

part of the eye	function
iris	*controls the amount of light entering the eye*
retina	
optic nerve	
cornea	

b The owl uses both eyes to see the same image.
Explain the advantage this type of vision gives the owl when it hunts.

_____

_____ **[1 mark]**

c Use ideas about **ciliary muscles**, **suspensory ligaments** and **lens** to
describe how the eye focuses on a distant object.

_____

_____

_____ **[3 marks]**

d Some people find it difficult to focus on distant objects because they are short-
sighted. Write down **two** different ways in which short sight can be corrected.

1 _____

2 _____ **[2 marks]**

2 The diagram shows the pathway taken by the impulse during an automatic reaction.

a Finish labelling the diagram to show the neurones involved in this process.

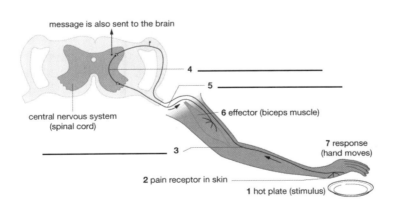

message is also sent to the brain

4 _____

5 _____

6 effector (biceps muscle)

central nervous system
(spinal cord)

7 response
(hand moves)

_____ 3

2 pain receptor in skin

1 hot plate (stimulus)

[3 marks]

b Describe how the impulse moves from one neurone to the next across the synapse.

_____

_____ **[2 marks]**

Drugs and you

1 a Different types of drugs have different effects. Finish the table by naming **one** example of each type of drug. Choose words from this list.

alcohol   **anabolic steroid**   **cannabis**   **caffeine**   **heroin**

type of drug	example
hallucinogen	
depressant	

[2 marks]

b Some people believe that personal use of cannabis should be allowed. Suggest **one** argument for and **one** argument against legal use of cannabis.

For _____

_____

Against _____

_____

[2 marks]

2 a Describe the effects of tobacco smoke on epithelial cells in the lining of the trachea.

_____

_____ [2 marks]

b Describe the effect of nicotine on synapses.

_____

_____ [2 marks]

3 a The picture shows drinks that contain one unit of alcohol. Matthew drinks two pints of beer and a glass of whisky. Jo drinks three glasses of wine and a glass of whisky. Who drinks the most units? Explain your answer.

half pint beer   single whisky   glass of wine

_____

_____ [2 marks]

b There are more alcohol-related injuries occurring on a Saturday compared to a Wednesday night. Suggest why.

_____

_____

_____ [2 marks]

Staying in balance

D–C

1 a The body can lose or gain heat. If the body gets too hot you can suffer from dehydration. Explain why.

_____

_____ [2 marks]

b When the body gets too cold the pulse rate slows. Name the condition the body suffers from when it gets too cold.

_____ [1 mark]

B–A*

c To prevent the body getting too warm vasodilation takes place. Explain what is meant by the term **vasodilation**.

_____

_____ [2 marks]

d Controlling body temperature involves negative feedback. Explain why.

_____

_____ [2 marks]

D–C

2 Blood sugar level is controlled by the hormone insulin.

| Sugar enters blood from small intestine | → | Blood sugar level rises | → | Insulin released into blood | → | Blood sugar level falls |

a Some people do not make enough insulin. Write down the name of the condition they suffer from.

_____ [1 mark]

b Suggest **one** way such people can control their blood sugar level.

_____ [1 mark]

B–A*

3 This diagram shows the menstrual cycle.

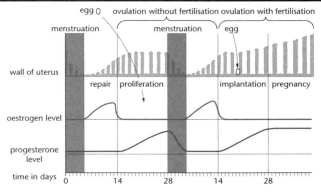

a Describe the effect of oestrogen on the lining of the uterus.

_____ [1 mark]

b Describe the effect of fertilisation on progesterone levels.

_____ [1 mark]

c Ovulation does not take place in some women. Suggest **one** way in which they can be treated.

_____ [1 mark]

Gene control

1 Finish the sentences about chromosomes.

Chromosomes are found in the _____ of the cell. They carry instructions called

_____. The chromosomes are made of a chemical called _____.　　[3 marks]

2 The squirrel developed from a fertilised egg. The fertilised egg is made up of 20 chromosomes.

a　How many chromosomes are in the egg before it is fertilised by a sperm?

_____　[1 mark]

b　How many chromosomes are in one cell from the squirrel's ear?

_____　[1 mark]

c　The nucleus of a human sperm is different from the nucleus of a squirrel sperm. Describe how they are different.

_____　[1 mark]

d　The cells in the squirrel's ears contain a code to make insulin but only the squirrel's liver cells make insulin. Explain why.

_____

_____　[2 marks]

3 Chromosomes contain special chemicals called bases.

a　How many different bases are there in a chromosome?

_____　[1 mark]

b　If a section of chromosome contained 30 **T** bases how many **A** bases will it contain? Explain your answer.

_____

_____　[2 marks]

c　This diagram shows how DNA controls eye colour. Changing the DNA code would change the eye colour. Explain how.

_____

_____

_____

_____

_____

_____

[3 marks]

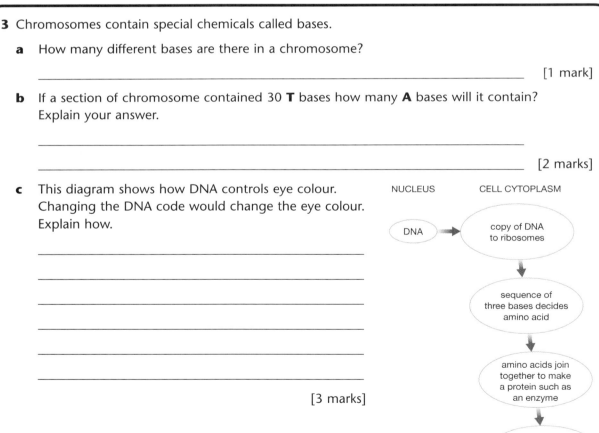

NUCLEUS　　　CELL CYTOPLASM

DNA → copy of DNA to ribosomes

sequence of three bases decides amino acid

amino acids join together to make a protein such as an enzyme

enzyme controls a reaction such as making an eye pigment

Who am I?

1 We inherit different characteristics from our parents' chromosomes.

a Finish the table to show how gender is inherited. [4 marks]

egg	sperm	fertilised egg	gender of child
X	X		
X	Y		

b The sperm determine gender. Use your knowledge of genetics to explain this.

_____

_____

_____ [2 marks]

2 Mutations are changes to genes. Mutations can occur spontaneously or they can be caused by some other factors.

a Write down **one** factor that can cause mutations.

_____ [1 mark]

b Haemophilia is a condition caused by a gene mutation. A protein needed to clot the blood cannot be made by the body. Explain why the mutation stops the protein being made.

_____

_____ [2 marks]

3 a The diagram shows a breeding experiment.

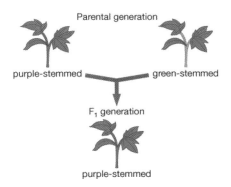

Parental generation

purple-stemmed — green-stemmed

F₁ generation

purple-stemmed

Which characteristic is dominant? Explain your answer.

_____ [1 mark]

b Cystic fibrosis is an inherited condition. Sam has cystic fibrosis but his parents are normal.

i Finish the genetic diagram to show how Sam inherited cystic fibrosis.

ii Put a (ring) around Sam's genotype on the diagram.

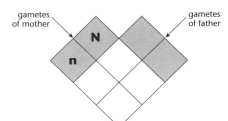

gametes of mother   gametes of father

N

n

[2 marks]

B1 Revision checklist

- I can explain the possible consequences of having high blood pressure. ☐

- I can state the word equation for respiration. ☐

- I know that religion, personal choice and medical issues can influence diet. ☐

- I can calculate BMI and RDA. ☐

- I can describe how white blood cells defend against pathogens. ☐

- I can explain the difference between passive and active immunity. ☐

- I can describe the functions of the main parts of the eye. ☐

- I know the classification and effects of the different types of drugs. ☐

- I can interpret data on the effects of smoking. ☐

- I can describe vasodilation and vasoconstriction. ☐

- I know that sex hormones can be used as contraceptives and for fertility treatment. ☐

- I know that chromosomes are made of DNA. ☐

- I can carry out a genetic cross to predict the possibilities of inherited disorders passing to offspring. ☐

- I know that inherited characteristics can be dominant or recessive. ☐

Cooking

D–C

1 Some foods need to be cooked. Explain why.

_____

_____ [2 marks]

D–C

2 a Write down **two** good sources of carbohydrate.

_____ [2 marks]

b Write down **two** good sources of protein.

_____ [2 marks]

B–A*

c Protein molecules change when they are cooked. Explain why this is important.

_____

_____

_____ [3 marks]

D–C

3 a Baking powder is a chemical called **sodium hydrogencarbonate**. When it is heated it **decomposes** to give sodium carbonate, carbon dioxide and water.

  i Write down the word equation for the reaction.

_____

_____ [2 marks]

  ii Write down the **reactants** of the reaction.

_____ [1 mark]

  iii Write down the **products** of the reaction.

_____ [1 mark]

B–A*

b The formula for sodium hydrogencarbonate is $NaHCO_3$. Write down the balanced symbol equation for this reaction.

_____

_____ [3 marks]

D–C

4 The chemical test for carbon dioxide is to pass it through limewater. It will turn the

limewater from _____

to _____ .

carbon dioxide ⟶

delivery tube

limewater

[2 marks]

Food additives

1 a Why are antioxidants added to tinned fruit?

_____ [1 mark]

b Ascorbic acid (vitamin C) is used as an antioxidant in which foods?

_____

_____ [2 marks]

c Information about food is given on food labels. How much energy does this food provide?

_____

Energy	56 J
Sodium	0.02 mg
Magnesium	Trace

[1 mark]

D–C

2 a Why is food packaging used?

_____ [1 mark]

b What is active packaging?

_____ [1 mark]

c Why is active packaging beginning to be used?

_____ [1 mark]

d It uses a **polymer**, what else is needed to make it work?

_____ [1 mark]

e Intelligent packaging works by including **indicators** on packages. Explain how these work.

_____ [1 mark]

D–C

B–A*

3 a Look at the diagram.
It is a detergent molecules made up of two parts, a head and a tail. Describe how the detergent works on removing grease in water. Use the diagram to help you.

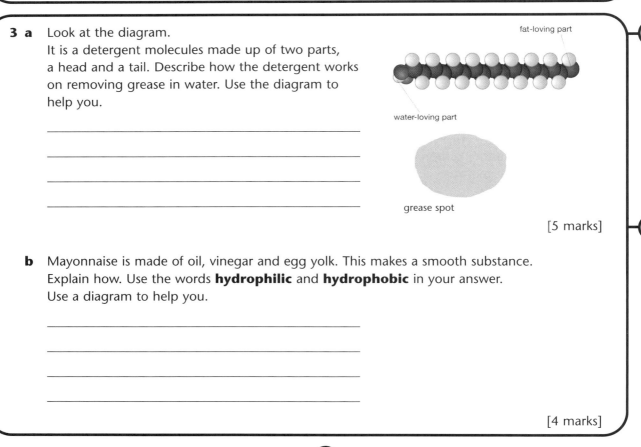

fat-loving part

water-loving part

grease spot

_____

_____

_____

_____

[5 marks]

D–C

B–A*

b Mayonnaise is made of oil, vinegar and egg yolk. This makes a smooth substance. Explain how. Use the words **hydrophilic** and **hydrophobic** in your answer. Use a diagram to help you.

_____

_____

_____

_____

_____

[4 marks]

Smells

1 a To make a perfume alcohol is mixed with an acid to make an ester.

    i Write down a word equation for this reaction.

_____

[2 marks]

    ii Look at the diagram. Label the alcohol and acid.   [1 mark]

    iii Label the condenser.   [1 mark]

    iv What is happening at **X**?

X—

_____

_____

[1 mark]

    v Why is the condenser used?

_____

_____

_____

[1 mark]

2 A good perfume needs to have several properties. These are listed in the boxes. Draw a **straight** line to match the **best** reason to the property needed.

evaporates easily	it can be put directly on the skin
non-toxic	its particles can reach the nose
insoluble in water	it does not poison people
does not irritate the skin	it cannot be washed off easily

[3 marks]

3 a A solute and a solvent that do not separate is a _____ .   [1 mark]

  b Esters are used as perfumes and _____ .   [1 mark]

4 a If a liquid evaporates easily then the substance is volatile. Explain how this happens. Use ideas about forces between particles in your answer.

_____

_____

_____

_____

[4 marks]

  b Water does not dissolve nail varnish. Use the diagram to help explain why.
Use ideas about the forces of attraction between molecules in your answer.

nail varnish molecules

water molecules

_____

_____

_____

_____

[2 marks]

Making crude oil useful

1 a All the oils of crude oil are **hydrocarbons**. What is a hydrocarbon?

_____

_____ [2 marks]

D–C

b The hydrocarbons are separated by **fractional distillation**.

  i Label the diagram A where the crude oil is heated. [1 mark]

  ii Label the diagram B where the fraction bitumen exits from. [1 mark]

  iii Label the diagram C at the coldest part. [1 mark]

  iv Which fraction 'exits' from the coldest part?

  _____ [1 mark]

c Explain why crude oil can be separated by fractional distillation.
Use ideas about chain length, forces of attraction and boiling points in your answer.

_____

_____

_____

_____

_____ [5 marks]

B–A*

2 Explain how damage is caused to the environment if oil tankers are damaged.

_____

_____ [2 marks]

D–C

3 a Cracking breaks down long-chain molecules called **alkanes**. They have a general formula of C_nH_{2n+2}. What is the molecular formula of heptane which has 7 carbon atoms?

_____ [1 mark]

D–C

b When a large alkane is cracked it becomes a smaller alkane and an alkene. Explain why an **alkene** is a different type of hydrocarbon to an alkane.

_____ [1 mark]

c What are alkenes useful for making? _____ [1 mark]

d A country produces 25% more than its demand of heavy oil from crude oil distillation. However, its supply of petrol from the distillation is only 68% of its need. Suggest how they could solve this problem other than by importing petrol from elsewhere.

_____

_____

_____ [3 marks]

B–A*

Making polymers

1 a Which molecule is a polymer? Put a (ring) around **A**, **B**, **C** or **D**.

A

$$H-\underset{\underset{H}{|}}{\overset{\overset{H}{|}}{C}}-\underset{\underset{H}{|}}{\overset{\overset{H}{|}}{C}}-OH$$

B

$$H-\underset{\underset{H}{|}}{\overset{\overset{H}{|}}{C}}-\underset{\underset{H}{|}}{\overset{\overset{H}{|}}{C}}-\underset{\underset{H}{|}}{\overset{\overset{H}{|}}{C}}-H$$

C

$$\left[\begin{array}{cccccc} | & | & | & | & | & | \\ C & C & C & C & C & C \\ | & | & | & | & | & | \end{array}\right]_n$$

D

$$C-\overset{\overset{H}{|}}{C}=\overset{\overset{H}{|}}{C}-Cl$$

[1 mark]

b Write down **two** conditions needed for polymerisation.

_____

_____ [2 marks]

c What does a monomer need in its structure to undergo addition polymerisation?

_____ [1 mark]

d Construct the **displayed formula** of the addition polymer from this monomer. [2 marks]

$$\underset{\underset{Cl}{|}\ \ \underset{H}{|}}{\overset{\overset{H}{|}\ \ \overset{H}{|}}{C=C}}$$

e Draw the monomer that makes this addition polymer. [2 marks]

$$\left[\begin{array}{cccccc} H & CH_3 & H & CH_3 & H & CH_3 \\ C & C & C & C & C & C \\ H & H & H & H & H & H \end{array}\right]_n$$

f Explain how addition polymerisation takes place.

_____

_____

_____

_____ [4 marks]

2

a i Butanol, C_4H_9OH $H-\underset{\underset{H}{|}}{\overset{\overset{H}{|}}{C}}-\underset{\underset{H}{|}}{\overset{\overset{H}{|}}{C}}-\underset{\underset{H}{|}}{\overset{\overset{H}{|}}{C}}-\underset{\underset{H}{|}}{\overset{\overset{H}{|}}{C}}-OH$ is **not** a hydrocarbon. Explain why.

_____ [1 mark]

ii Butene is an alkene. $H-\underset{\underset{H}{|}}{\overset{\overset{H}{|}}{C}}-\underset{\underset{H}{|}}{\overset{\overset{H}{|}}{C}}-\overset{\overset{H}{|}}{C}=\overset{\overset{H}{|}}{C}-H$

Explain how you know.

_____ [1 mark]

iii Butene is a **monomer**. What is **polybutene**?

_____ [1 mark]

b i Bromine solution is used to test for unsaturation. It is an orange solution. When an alkene is added the orange solution turns colourless. Explain why.

_____ [1 mark]

ii When bromine solution is added to an alkane, what do you see?

_____ [1 mark]

Designer polymers

1 a Polymers are better than other materials for some uses. Give **one** example of the use of a polymer in healthcare that is better than the material used before.

_____

_____ [2 marks]

D–C

b i Gore-Tex® type materials are used to make clothing waterproof and breathable. The inner layer of the clothing is made from expanded PTFE (polytetrafluoroethene) which is **hydrophobic**. Explain what hydrophobic means.

_____

_____ [2 marks]

B–A*

ii The PTFE is expanded to form a **microporous membrane**. Explain how this makes a breathable membrane.

_____

_____ [2 marks]

2 a Scientists are developing addition polymers that are **biodegradable**. Explain why.

_____

_____ [2 marks]

D–C

b Suggest a use for a biodegradable plastic.

_____ [1 mark]

c Disposing of non-biodegradable polymers causes problems. Explain the problems for each of the ways of disposing.

Landfill sites _____

Burning waste plastic _____

Recycling _____ [3 marks]

3 Look at the diagram of polymer chains.

B–A*

a i Label the **intramolecular covalent** bonds. [1 mark]

ii Label the **intermolecular** forces of attraction. [1 mark]

b Some plastics have low melting points and can be stretched easily whereas other plastics have high melting points and cannot be stretched easily. They are rigid. Explain why. Use ideas about forces between chains in your answer. You may use a diagram to help you.

_____

_____

_____

_____ [4 marks]

Using carbon fuels

D–C

1 a i Look at the table. Which fuel produces more acid fumes?

characteristic	coal	petrol
energy value	high	high
availability	good	good
storage	bulky and dirty	volatile
toxicity	produces acid fumes	produces less acid fumes
pollution caused	acid rain, carbon dioxide and soot	carbon dioxide, nitrous oxides
ease of use	easy in power stations	easy in engines

_____ [1 mark]

ii Give **two** advantages of using either coal or petrol for heating.

_____

_____ [2 marks]

B–A*

b Explain why the use of petrol and diesel in transport is contributing to a global problem.

_____

_____

_____

_____ [4 marks]

D–C

2 a Write down a **word equation** for a hydrocarbon fuel burning in air.

_____ [1 mark]

b Two products are made in the complete combustion of a fuel. These can be tested in the laboratory.
i Limewater is used to test for one product. Which one?

_____ [1 mark]

ii How is the other product tested?

_____ [1 mark]

c Complete combustion is better than incomplete combustion. Explain why.

_____

_____

_____ [3 marks]

d Why should a room be well ventilated and a heater regularly checked?

_____

_____ [2 marks]

B–A*

e Write a balanced equation for the complete combustion of propane, C_3H_8.

_____ [2 marks]

Energy

1 a Use the words **exothermic** and **endothermic** correctly in these sentences.

When energy is transferred **out** to the surroundings in a chemical reaction it is an

_____ reaction (energy is released).

When energy is taken in from the surroundings in a chemical reaction it is an

_____ reaction (absorbs energy).

An_____ reaction is shown by a temperature **increase**.

Burning magnesium is an example of an _____ reaction. [4 marks]

b Is bond breaking an exothermic or endothermic process?

_____ [1 mark]

c Burning methane is an exothermic reaction. Explain why, using ideas about bond breaking and bond making.

_____

_____

_____ [3 marks]

---

2 The flame of a Bunsen burner changes colour depending on the amount of

oxygen that it burns in. The flame is _____ when the gas burns in plenty

of oxygen. This is because there is _____ combustion. The flame

is _____ when the gas burns in limited oxygen. This is because

there is _____ combustion. [4 marks]

---

3 a Describe how you would design your own experiment to compare the energy transferred by two different fuels, using a beaker of water and a thermometer. Write down the ways to make the tests fair.

_____

_____

_____

_____

_____ [5 marks]

b Todd and Terri calculate the amount of energy transferred during a reaction. They use a spirit burner to burn the fuel to heat 100 g of water in a copper **calorimeter**. They measure a temperature change in the water of 50 °C. The mass of fuel they burn is 4.00 g.

i How do they make the tests reliable?

_____ [1 mark]

ii Show how they calculate the energy released per gram. (The specific heat capacity of water is 4.2 J/g/°C)

[5 marks]

C1 Revision checklist

- I know that cooking food is a chemical change as a new substance is made and it is an irreversible reaction. ☐

- I know that protein molecules in eggs and meat change shape when the food is cooked. ☐

- I know that when the shape of a protein changes it is called denaturing. ☐

- I know that emulsifiers are molecules that have a water-loving part and an oil- or fat-loving part. ☐

- I know that alcohols react with acids to make an ester and water. ☐

- I know that a solute is the substance dissolved in a solvent to make a solution that does not separate. ☐

- I know that crude oil is a non-renewable fossil fuel, which is a mixture of many hydrocarbons. ☐

- I know that petrol is a crude oil fraction with a low boiling point, which exits at the top of the fractional distillation tower. ☐

- I know that polymerisation is a process which requires high pressure and a catalyst. ☐

- I know that a hydrocarbon is a compound formed between carbon atoms and hydrogen atoms only. ☐

- I know that alkenes are hydrocarbons with one or more double bonds between carbon atoms. ☐

- I know that complete combustion of a hydrocarbon fuel makes carbon dioxide and water only. ☐

- I know that an exothermic reaction is one where energy is released into the surroundings. ☐

- I can work out that: energy transferred = mass of water × 4.2 × temperature change. ☐

Heating houses

Grades

1 Kelly opens the front door on a very cold morning. Her mother complains that the house is getting cold.
Use your ideas about energy flow to explain why the house gets cold.

_____

_____ [2 marks]

D–C

2 A police helicopter uses a thermal imaging camera to take a picture at night. It is looking for a car that has recently been abandoned in a field after a high speed chase. How does the thermogram help locate the car?

_____

_____

_____ [3 marks]

B–A*

3 a Finish the sentence.
The energy needed to raise the temperature of 1 kg of a material by 1 °C is known as the _____. [1 mark]

D–C

b The specific heat capacity of seawater is 3900 J/kg °C. How much energy is needed to heat 500 g of seawater from 20 °C to 90 °C?
Show how you work out your answer.

B–A*

_____ [4 marks]

4 a What physical quantity is measured in units of J/kg?

_____ [1 mark]

D–C

b When iron changes from solid to liquid, energy is transferred but there is no temperature change until all of the solid has changed into liquid. Use your ideas about molecular structure to explain why there is no change in temperature at iron's melting point.

_____

_____

_____ [2 marks]

B–A*

Keeping homes warm

1 Dan wants to reduce the heating bills for his old house. He decides to insulate his loft first and then replace his windows with double glazing.

a Dan spends £120 on loft insulation. He is told that this will reduce his heating bills by £40 per year. Calculate the payback time for loft insulation.
Show how you work out your answer.

_____ [2 marks]

b Why does Dan decide to insulate the loft before replacing his windows?

_____

_____ [1 mark]

c Dan heats his house with coal fires. He is told that his fires are **32% efficient**. Explain what is meant by 32% efficient.

_____

_____ [2 marks]

d Dan pays £6.50 for a 25 kg bag of coal. How much of that money is **usefully** used in heating his house?

_____ [2 marks]

e Why are coal fires so inefficient?

_____

_____ [1 mark]

f What types of materials transfer energy?

i by conduction

_____ [1 mark]

ii by convection

_____ [1 mark]

iii by radiation

_____ [1 mark]

How insulation works

1 a The diagram shows a section through a double-glazed window. Michael says that it is just as effective to use a piece of glass twice the thickness. Use your ideas about energy transfer to explain why double glazing is better.

_____

_____

_____

_____

_____

space filled with air or argon, or has a vacuum

[3 marks]

b New homes are built with insulation blocks in the cavity between the inner and outer walls. The blocks have shiny foil on both sides.

block wall

exterior brick or stone finish

solid foam board

i Explain how the insulation blocks reduce energy transfer by conduction and convection.

_____

_____

_____

_____

[4 marks]

ii Explain how the shiny foil helps to keep a home warmer in winter and cooler in summer.

_____

_____

[2 marks]

2 a A brick in a wall is a better conductor of heat than air in the cavity between the walls.
Use your ideas about particles and kinetic energy to explain this.

_____

_____

_____

[3 marks]

b Hot air rises. Explain why.

_____

_____

_____

[3 marks]

Cooking with waves

1 a Why do microwave ovens take less time to cook food than normal ovens?

_____ [1 mark]

b Microwaves are suitable to communicate with spacecraft thousands of kilometres away while sometimes a mobile phone cannot receive a signal just a few kilometres from the nearest transmitter. Why do microwave signals seem to work better in space than they do on Earth?

_____

_____ [2 marks]

c Dave says that microwave signals 'bounce' off satellites. Sally says Dave is wrong. What happens to a microwave signal when it is received by a satellite?

_____

_____ [2 marks]

2 The diagram shows the electromagnetic spectrum.

a Which part of the electromagnetic spectrum transfers the most energy? _____

[1 mark]

b An electric iron transfers less energy than the element of an electric fire. How does the wavelength of radiation from the iron differ from the wavelength of radiation from the element?

_____

_____ [1 mark]

3 The diagram shows a transmitter on top of a hill. It transmits microwave mobile phone signals as well as radio and television signals.

The house, behind the other hill, can receive radio and television signals but there is no mobile phone reception. Use your knowledge of microwave properties to explain why.

_____

_____

_____ [2 marks]

Infrared signals

1 a The diagram shows a signal displayed on an oscilloscope. What type of signal is it? Put a ⟨ring⟩ around the correct answer.

background	**digital**	**on/off**	**radial**	**square**	[1 mark]

b Draw a diagram to represent an **analogue** signal.

[1 mark]

c Computers do not use analogue signals to transmit information. Explain why.

_____

_____

_____

_____ [3 marks]

2 a The diagrams show three rays of light travelling from water into air. The three angles of incidence are (x) smaller than the critical angle (y) equal to the critical angle (z) larger than the critical angle.

(x)   (y)   (z)

i Finish the diagrams to show what happens to the rays of light as they continue towards the water/air boundary and after they meet the water/air boundary. [5 marks]

ii Label the critical angle on the correct diagram with the letter **c**. [1 mark]

b Adam's doctor uses an endoscope to look at the inside of his stomach.
  i Write down the advantages of using an endoscope.

_____

_____ [2 marks]

  ii Explain how an endoscope allows the doctor to see inside Adam's stomach.

_____

_____

_____ [3 marks]

Wireless signals

1 Radio waves are refracted in the upper atmosphere. How is the amount of refraction affected if the frequency of the radio wave is decreased?

_____ [1 mark]

2 Luke and Jessica are walking in the mountains. They are both listening to their personal radios.
Luke is listening to Radio 1 on FM and Jessica is listening to Radio 4 on long wave. Both radio stations are broadcasting from the same aerial. Luke complains that he keeps losing the signal but Jessica does not have this problem.
Use your ideas about waves to explain why Luke has problems with reception but Jessica does not.

_____

_____

_____ [3 marks]

3 A satellite is in orbit, 36 000 km above the equator. It takes 24 hours to orbit Earth. Microwaves travel at the speed of light (300 000 km/s).

a Why does a communications satellite take 24 hours to orbit Earth?

_____ [1 mark]

b How long does it takes for a microwave signal to travel from Earth, to the satellite and back to Earth? Put a ⟨ring⟩ around the correct answer.

      0.12 s   **0.24 s**   **1.2 s**   **2.4 s**   [1 mark]

4 a Ella listens to her favourite radio station. Every so often, she notices that she can hear a foreign radio station as well. Put ticks (✓) in the boxes next to the **two** statements that explain why this happens.

The foreign radio station is broadcasting on the same frequency.	
The foreign radio station is broadcasting with a more powerful transmitter.	
The radio waves travel further because of weather conditions.	
Ella's radio needs new batteries.	

[2 marks]

b Why does the microwave beam sent by a transmitting aerial towards a satellite in orbit have to be **focused**?

_____ [1 mark]

Light

---

1 The diagram shows a transverse wave.

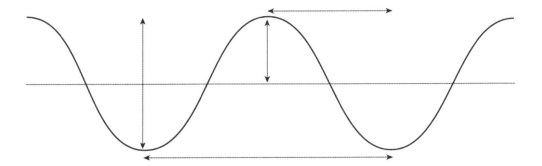

a Write the letter **A** next to the arrow which shows the amplitude of the wave. [1 mark]

b Write the letter **W** next to the arrow which shows the wavelength of the wave. [1 mark]

c What is meant by the **frequency** of a wave?

_____

_____ [1 mark]

D–C

---

2 a Why was it necessary for Samuel Morse to devise a code of dots and dashes?

_____ [1 mark]

b Lisa is using a signalling lamp to send a message to Robbie. Write down **one** advantage and **one** disadvantage of using a signalling lamp.

Advantage _____

_____

Disadvantage _____

_____ [2 marks]

c Lisa's science teacher shines a laser light onto a screen. It is brighter than the white light from the laboratory lights. Use your ideas about frequency to explain the difference between laser light and white light. Draw diagrams to help you answer the question.

_____

_____

_____

_____

[4 marks]

D–C

B–A*

Stable Earth

D–C

1 a **P** waves and **S** waves are two of the waves which travel through the Earth after an earthquake. Put a tick (✓) in the box or boxes to correctly describe each wave. The first one has been done for you.

description	P wave	S wave
pressure wave	✓	
transverse wave		
longitudinal wave		
travels through solid		
travels through liquid		

[5 marks]

B–A*

b How do scientists use the properties of **P** waves and **S** waves to find out the **size** of the Earth's core?

_____

_____ [2 marks]

c How do scientists use the properties of **P** waves and **S** waves to find out the **structure** of the Earth's core?

_____

_____ [2 marks]

D–C

2 a Why does dust from factory chimneys cause warming of the Earth?

_____ [1 mark]

B–A*

b What effect do CFCs from aerosols and refrigerators have on the ozone layer?

_____

_____ [2 marks]

c Why is it beneficial to Earth to have a layer of ozone surrounding the planet?

_____ [1 mark]

D–C

d Leah wants to sunbathe and get a good tan.

i What causes Leah's skin to tan?

_____

_____ [2 marks]

ii She uses a sun screen with **SPF 30**. What does SPF 30 mean?

_____

_____ [2 marks]

P1 Revision checklist

- I can explain the difference between temperature and heat. ☐

- I can explain what is meant by specific heat capacity and specific latent heat. ☐

- I can perform energy calculations involving specific heat capacity and specific latent heat. ☐

- I can explain why different forms of domestic insulation work in terms of energy transfer. ☐

- I can calculate energy efficiency. ☐

- I can explain why infrared radiation is used for cooking. ☐

- I can explain why microwaves are used for cooking and for communication. ☐

- I can explain the advantages of using digital signals instead of analogue signals. ☐

- I can describe total internal reflection and its use in optical fibres, including the endoscope. ☐

- I can explain how wireless signals are used for global communication. ☐

- I can explain why there is sometimes interference with microwave and radio signals. ☐

- I know the properties of a transverse wave and how light is used as a transverse wave. ☐

- I can describe earthquake waves and how they can tell us about the structure of the Earth. ☐

- I know some of the effects of natural events and human activity on weather patterns. ☐

Ecology in our school grounds

1 a Clown fish are found in the coral reefs of the Pacific Ocean.

   i Suggest a reason why the clown fish are not found in British lakes.

_____ [1 mark]

   ii Suggest a reason why there are many undiscovered species in the Pacific Ocean.

_____ [1 mark]

b Wheat fields are artificial ecosystems and woods are natural ecosystems.
Woods tend to have a higher biodiversity.

   i What is meant by the term **biodiversity**?

_____ [1 mark]

   ii Use your knowledge about invertebrates and vertebrates to explain how the use of pesticides could reduce the biodiversity of the wheat field.

_____

_____ [2 marks]

2 The following formula is used to estimate a population.

$$\frac{\dfrac{\text{number of animals caught}}{\text{first time}} \times \dfrac{\text{number of animals caught}}{\text{second time}}}{\text{number of marked animals caught second time}} = \text{population}$$

Researchers want to know the number of voles living in a wood. They set 30 traps in a small area of the wood and catch 20 voles. They mark the voles and then release them back into the wood. A week later they set more traps and catch 10 voles, five of them are marked.

a Estimate the population of voles in the wood.

_____

_____ [2 marks]

b The researchers are not convinced that their estimate is correct.

   i Suggest **two** ways in which they can improve their method.

_____

_____ [2 marks]

   ii Explain why your suggestions would make the estimate more accurate.

_____

_____ [2 marks]

Grouping organisms

1 a The picture shows a penguin.

    i Finish this sentence.

    The penguin belongs to the vertebrate group

    called _____ .

    [1 mark]

    ii Salmon belong to the vertebrate group called fish because they have scales. Explain why the penguin belongs to the group you have chosen.

    _____

    _____

    [2 marks]

b Explain why spiders are invertebrates.

    _____

    [1 mark]

c This table compares animals and plants. Finish the table to describe animals.

	food	shape	movement
plants	*make own*	*spread out*	*stay in one place*
animals			

[3 marks]

D–C

2 A Zorse is a cross between two different species, a zebra and a horse.

a What name is used to describe a cross between two different species?

    _____

    [1 mark]

b Explain why the Zorse is difficult to classify.

    _____

    [1 mark]

B–A*

3 The lion and tiger are different species.

a What is meant by the term **species**?

    _____

    _____

    [2 marks]

b Lions and tigers belong to the same family of cats. This table shows the Latin names of different cats.

common name	Latin name
bobcat	*Felix rufus*
cheetah	*Acinonyx jubatus*
lion	*Panthera leo*
ocelot	*Felix pardalis*

[3 marks]

    i Two of these cats are more closely related than the others. Write down the **common** names of these **two** cats.

    _____ and _____

    [1 mark]

    ii What is the reason for your answer to part **b i**?

    _____

    [1 mark]

D–C

The food factory

B–A*

1 Plants make glucose ($C_6H_{12}O_6$) by a process called photosynthesis.

D–C

   a Finish the balanced symbol equation for photosynthesis.

      _____ + _____ → $C_6H_{12}O_6$ + _____   [1 mark]

   b The products of photosynthesis have many uses. Finish the table to describe these uses. The first one has been done for you.

product of photosynthesis	use in the plant
glucose	*energy*
cellulose	
protein	
oil	

                                                       [3 marks]

D–C

2 Light can change the rate of photosynthesis.

   a Write down the names of **two other** factors that change the rate of photosynthesis.

     1 _____

     2 _____   [2 marks]

   b The graph shows the effect of light intensity on the rate of photosynthesis.

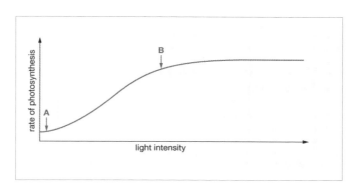

     i Describe the pattern shown in the graph.

     _____

     _____

     _____   [2 marks]

D–C

3 Glucose is a product of photosynthesis. The plant uses a process to release energy from glucose.

   a Write down the name of this process.

     _____   [1 mark]

   b Plants carry out this process 24 hours a day. Explain why.

     _____   [1 mark]

Compete or die

1 a The red and grey squirrels share the same ecological niche. What is meant by the term **ecological niche**?

_____   [1 mark]

B–A*

b The red squirrel is native to Britain. The grey squirrel was introduced to Britain about 130 years ago. During the last 100 years grey squirrel numbers have increased and red squirrel numbers have decreased.

D–C

i Suggest a reason for the change in numbers.

_____   [1 mark]

ii Grey squirrels have been removed from the island of Anglesey. Suggest the effect this might have on the red squirrel population.

_____   [1 mark]

2 This is a diagram of a food chain.

D–C

a In one year the population of shrews decreases.

i Describe the effect this would have on the badger population.

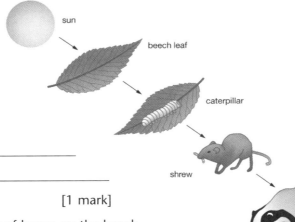

_____

_____

[1 mark]

ii In the same year the number of leaves on the beech tree decreases. Explain why.

_____

_____   [2 marks]

b The oxpecker feeds on the insects found on the buffalo's skin. The insects are called parasites.

i What is meant by the term **parasite**?

_____   [1 mark]

ii The oxpecker and the buffalo both benefit from the relationship. Write down the name given to this type of relationship.

_____   [1 mark]

3 Look at the graph. It shows the predator-prey relationship of lemmings and snowy owls.
Write about how and why the number of lemmings change as indicated in the graph.

B–A*

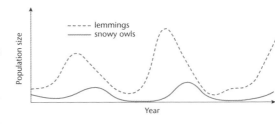

_____

_____

_____

_____   [3 marks]

Adapt to fit

1 a Look at the picture of a camel. The table shows how the camel is adapted to help it live in the desert. Finish the table by writing how these adaptations help it to survive. The first one has been done for you.

adaptation	why it helps the camel survive
large feet	*stop sinking into sand*
no fat on body, except in hump	
hair-lined nostrils	
higher body temperatures do not harm camel	

[3 marks]

b Polar bears are adapted to live in the cold. They have thick fur for insulation as this stops them losing too much body heat.

  i Write about other ways they are adapted to live in the cold.

_____

_____

_____

_____

[3 marks]

  ii Polar bears are not in found in the same habitat as brown bears. Explain why. Use ideas about competition in your answer.

_____

_____

[2 marks]

2 a The cactus has spines instead of leaves. Explain why.

_____

_____

[2 marks]

b Lilies are pollinated by insects. Write down **one** adaptation of insect-pollinated flowers.

_____

[1 mark]

c Grasses are pollinated by wind. Descibe **two** ways they are from different insect-pollinated plants.

_____

_____

[2 marks]

Survival of the fittest

1 a Here are four sentences (**A–D**) about how the fossil of a dinosaur is formed. They are in the wrong order. Write the letters in the boxes to show the right order. The first one has been done for you.

A The dinosaur's hard parts were replaced by minerals.

B The dinosaur died.

C The dinosaur's soft tissue rotted away.

D The dinosaur became covered by sediment.

B			

[2 marks]

D–C

b The fossil record shows how living things have changed over time. The fossil record is incomplete. Explain why.

_____

_____ [2 marks]

c Creationists believe that each living thing was created individually and did not evolve. Use your knowledge about the fossil record to suggest **one** argument for and **one** argument against this theory.

For _____

_____

Against _____

_____ [2 marks]

B–A*

2 The following article gives information on the superbug MSRA. Read the article carefully and use it to help you answer the questions.

MRSA Where did it come from?
MRSA evolved because of natural selection. There are lots of different strains of the bacteria. Each strain has slightly different DNA. The DNA is also constantly mutating as the bacteria reproduce. Some of these mutations will be more resistant to antibiotics than others.

When people take antibiotics, the less resistant strains die first, the more resistant strains are harder to destroy. If people stop taking the antibiotics too soon the resistant strains survive to reproduce and pass on their DNA. In this way, more and more strains evolve to be resistant to these new drugs.

a What is meant by the term **natural selection**?

_____ [1 mark]

D–C

b Use Darwin's theory of natural selection to explain how MRSA has evolved.

_____

_____ [2 marks]

B–A*

Population out of control?

1 a The rise in human population is causing an increase in the level of carbon dioxide in the air. Suggest **two** effects this increase may have on the environment.

1 _____

2 _____ [2 marks]

b The ozone layer in the Earth's atmosphere protects us from harmful ultraviolet rays. Chemicals are destroying the ozone layer.

i Write down the name of these chemicals.

_____ [1 mark]

ii The over-use of these chemicals has caused an increase in skin cancer. Suggest a reason why.

_____ [1 mark]

2 The graph shows the past, present and predicted future of the world human population.

a The human population is in the rapid growth stage. What is the name given to this stage of growth?

_____

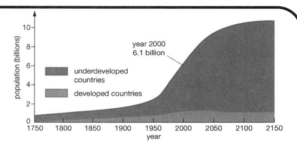

[1 mark]

b In developed countries, such as America, the population is constant. In underdeveloped countries, such as Africa, the population is still rising, yet underdeveloped countries cause less pollution. Suggest **two** reasons why.

1 _____

2 _____ [2 marks]

3 Scientists look for the water louse when they want to measure the level of pollution in water.

a Write down the name given to species that are used to measure levels of water pollution.

_____ [1 mark]

b This table shows the sensitivity of different animals to pollution.

animal	sensitivity to pollution	animal	sensitivity to pollution
stonefly larva	sensitive	freshwater mussel	semi-sensitive
water snipe fly	sensitive	damselfly larva	semi-sensitive
alderfly	sensitive	bloodworm	tolerates pollution
mayfly larva	semi-sensitive	rat-tailed maggot	tolerates pollution

A river sample contains mussels, damselfly larva and bloodworms, but no alderfly or stonefly larva. Use this information to explain why the river sample is polluted.

_____

_____ [2 marks]

Sustainability

1 This is a picture of the dodo. The dodo no longer exists.
Suggest **two** reasons why the dodo no longer exists.

1 _____

2 _____ [2 marks]

2 Some animals are close to extinction. They are called **endangered species**.
The panda is an endangered species.

a Pandas are kept in captivity. Suggest how this may help increase the panda population.

_____ [1 mark]

b Pandas live in a remote part of China. Their natural habitat is being destroyed.
Some people want to save the panda from extinction. Suggest **two** reasons why
saving the panda will help the people who live in the same habitat.

1 _____

2 _____ [2 marks]

3 a Some countries want to hunt whales for food. Suggest **one** argument for and
one argument against hunting whales.
For _____

_____

Against _____

_____ [2 marks]

b It is very difficult to stop people hunting whales.
Suggest **one** reason why.

_____ [1 mark]

4 a Sustainable resources are resources that should not run out. Fish are a sustainable
resource. The government has set fish quotas. Fishermen can only catch so many
fish at any one time. The fish quotas should help maintain the population of fish in
the sea. Explain why.

_____

_____ [2 marks]

b As the human population increases there will be more demand for energy
resources. The sustainable development of fast-growing willow trees could help
solve the problem. Explain why.

_____

_____ [2 marks]

B2 Revision checklist

- I know how to collect and use data to estimate a population. ☐

- I know how to use a key to identify plants and animals. ☐

- I know the characteristics of the different vertebrate groups. ☐

- I can state the word and formula equation for photosynthesis. ☐

- I can describe the effect of increased light, temperature and carbon dioxide on photosynthesis rate. ☐

- I can explain how similar organisms compete for the same ecological niche. ☐

- I can explain how the size of a predator population will affect the prey population. ☐

- I can explain how adaptations of organisms determine their distribution and abundance. ☐

- I can explain how camels, polar bears and cacti are adapted to their habitats. ☐

- I can describe how organisms became fossilised. ☐

- I know that developed countries have a greater effect on world pollution. ☐

- I can explain the effects of increased pollution on climate change, acid rain and the ozone layer. ☐

- I can describe ways in which animals become extinct. ☐

- I can explain sustainable development and describe how it may protect endangered species. ☐

Paints and pigments

1 a An **emulsion paint** is a water-based paint. Explain how it covers a surface.

_____

_____ [2 marks]

D–C

b Sam and Chris are discussing why paint is a colloid. Sam gives an explanation. He uses ideas about particle size and mixtures and dispersion. Write down what he says.

_____

_____

_____

_____ [4 marks]

B–A*

c The oil in oil paint is very sticky and takes a long time to harden. Explain how it hardens in two stages and what it forms.

_____

_____

_____ [3 marks]

2 a Thermochromic pigment changes colour at 45 °C. Write down **two** examples it is used for.

_____

_____ [2 marks]

D–C

b Most thermochromic pigments change from a particular colour to colourless. How do they ensure a greater range of colours is available?

_____ [1 mark]

B–A*

c Thermochromic paints come in a limited range of colours. If a green paint becomes yellow when heated explain what the mixture contains and how it changes colour.

_____

_____

_____

_____ [4 marks]

d A pigment that stores absorbed energy is called

_____ [1 mark]

D–C

e What does it release that energy as?

_____ [1 mark]

3 a Where are phosphorescent pigments used?

_____ [1 mark]

B–A*

b What have phosphorescent pigments replaced and why?

_____

_____ [2 marks]

Construction materials

D–C

1 a Put these materials into the order of hardness.

        granite   **limestone**   **marble**

Least hard _____ _____ _____ Hardest   [3 marks]

b Brick, concrete, steel, aluminium and glass are manufactured.
Finish the table to show the raw materials they come from.

building material	brick	cement	glass	iron	aluminium
raw material					

[5 marks]

B–A*

c **Igneous**, **sedimentary** and **metamorphic** rocks differ in hardness.

Which rock is usually the least hard? _____   [1 mark]

d Granite and marble are different types of rock. Their hardness is different. Compare how granite and marble are made and why their structures are different. Use **all** the words in this list.

crystals igneous interlocking heat pressure slowly metamorphic solidifies

granite	marble

[8 marks]

e Limestone is a **sedimentary rock**. Explain how limestone is made.

_____

_____   [3 marks]

D–C

2 a Calcium carbonate decomposes at a very high temperature. Write a word equation.

_____   [3 marks]

b Cement is made from limestone. Write down how.

_____   [2 marks]

B–A*

c Write down the symbol equation for the thermal decomposition of calcium carbonate.

_____   [3 marks]

d Reinforced concrete is a better construction material than non-reinforced concrete. Explain why. Use ideas about stretching and tension in concrete and strength under tension of steel.

_____

_____

_____

_____   [4 marks]

Does the Earth move?

1 a Are the tectonic plates that make up the Earth's crust less dense or more dense than the mantle?

_____ [1 mark]

D–C

b Write down the **two** kinds of tectonic plate.

_____ and _____ [2 marks]

c The mantle is always solid, but at greater depths it is more like Plasticine, which can 'flow'. The **tectonic plates** move very slowly on this mantle. They can move in different ways. Explain how, using diagrams to help you explain.

B–A*

[5 marks]

d Explain how subduction occurs.

_____

_____

_____

_____

_____ [6 marks]

e Write down **two** pieces of evidence used to develop the theory of plate tectonics. Use ideas about continental drift and mid-ocean ridges.

_____

_____

_____ [4 marks]

2 a Magma can rise through the Earth's crust. Explain why. [1 mark]

D–C

_____

b Magma cools and solidifies into igneous rock either after it comes out of a volcano as lava, or before it even gets to the surface. By looking at crystals of igneous rock, geologists can tell how quickly the rock cooled. Fill in the **two** boxes with an explanation and an example.

basalt	cools rapidly		cools slowly	granite

small crystals

large crystals

[4 marks]

c i There are two types of magma, iron-rich and silica-rich. Which is the magma that is less runny and produces volcanoes that may erupt explosively?

B–A*

_____ [1 mark]

ii Explain what it produces.

_____

_____ [2 marks]

Metals and alloys

D–C

1 a Copper used for recycling has to be sorted carefully so that valuable 'pure' copper scrap is not mixed with less pure scrap. When impure copper is used to make alloys what must happen first?

_____ [1 mark]

b If the scrap copper is very impure what must be done before it is used again?

_____ [1 mark]

D–C

2 Impure copper can be purified in the laboratory using an electrolysis cell.

a What is the **anode** made from?

_____ [1 mark]

b What happens at the **cathode**?

_____ [1 mark]

B–A*

c What happens to the **anode**?

_____ [1 mark]

d What happens to the thickness of the cathode?

_____ [1 mark]

e What is the impure copper called?

_____ [1 mark]

f What happens to the impurities from the copper anode?

_____ [1 mark]

D–C

3 a Most metals form alloys. Draw a **straight** line to match the metals to the alloy.

| amalgam | | contains copper and zinc |
| solder | | contains mercury |
| brass | | contains lead and tin | [2 marks]

b Alloys are often more useful than the original metals, though nowadays pure copper is more important than bronze or brass. Why are vast amounts turned into electric wire?

_____ [1 mark]

B–A*

c Write down **two** properties of smart alloys.

_____
_____ [2 marks]

d Write down **two** new ways of using smart alloys.

_____
_____ [2 marks]

e **Nitinol** is a smart alloy. Which **two** metals is it made from?

_____
_____ [2 marks]

Cars for scrap

1 a In winter, icy roads are treated with salt. Why is this a problem for steel car bodies?

_____ [1 mark]

b Aluminium does not corrode in moist air. Explain why.

_____ [1 mark]

c Rust is an oxide layer but it does not protect the rest of the iron. Explain why.

_____ [1 mark]

d Rusting is a **chemical reaction** between iron and oxygen to make an oxide. The chemical name for rust is **hydrated iron(III) oxide**.
Write down the word equation for rusting.

_____ [3 marks]

2 a Steel is an alloy made of _____ and _____. [2 marks]

b Write down **two** advantages of steel over iron.

_____ [2 marks]

c Steel and aluminium can both be used to make car bodies but each material has its own advantages. Write down **two** advantages of cars made from aluminium and **one** disadvantage.

_____

_____

_____ [3 marks]

d Finish this table. Write down the reasons why these materials are used.

material and its use	reasons material is used
aluminium in car bodies and wheel hubs	
copper in electrical wires	
plastic in dashboards, dials, bumpers	
pvc in metal wire coverings	
plastic/glass composite in windscreens	
fibre in seats	

[6 marks]

3 Write down **three** benefits that recycling metals and the other materials of a car have on the environment.

_____

_____

_____ [3 marks]

D–C

B–A*

D–C

B–A*

D–C

1 a i Label the pie chart with the four main gases of the air in the correct section.

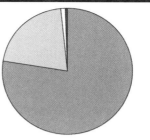

[4 marks]

ii Mark in the percentages of the gases.

[4 marks]

iii These percentages do not change very much because there is a balance between three of the processes that use up or make carbon dioxide and use or make oxygen. Explain how the balance is maintained.

_____

_____

_____

_____

[7 marks]

2 a Over the last few centuries the percentage of carbon dioxide in air has increased slightly. This is due to a number of factors. Explain these factors.

increased energy usage	
increased population	
deforestation	

[3 marks]

b Explain **one** theory of how the Earth's atmosphere has evolved. Include in your answer
i the original gases of the atmosphere
ii how nitrogen levels increased
iii how the levels of carbon dioxide and oxygen evolved to those which they are today.

_____

_____

_____

[4 marks]

3 a Finish the table to describe the origin of these atmospheric pollutants.

pollutant	carbon monoxide	oxides of nitrogen	sulphur dioxide
origin of pollutant			

[3 marks]

b Which metal catalyst do catalytic converters contain?

_____

[1 mark]

c i A reaction between nitric oxide and carbon monoxide takes place on the surface of the catalyst. The reaction forms nitrogen and carbon dioxide. Why is it important that these two gases are made?

_____

[1 mark]

ii Write down the word equation for this reaction.

_____

[2 marks]

iii Write down the balanced symbol equation for this reaction.

_____

[3 marks]

Faster or slower (1)

1 a Look at **Graph A**. It shows the reaction between magnesium and acid at 20 °C.

Graph A

D–C

  i If the reaction takes place at a higher temperature mark the reaction line that you would expect on the graph.

[2 marks]

  ii Which graph has the steeper gradient?

_____

_____

[1 mark]

  iii The reaction rate increases at higher temperatures. Explain why. Use ideas about particles in your answer.

_____

_____

[4 marks]

b The rate of a reaction is not determined by the number of collisions. Explain what it is determined by.

B–A*

_____

_____

[2 marks]

c Explain the increase in reaction rate in terms of particle collisions.

_____

[1 mark]

2 Look at **Graph B**. It shows the reaction between magnesium and acid of different concentrations.

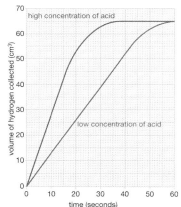

Graph B

D–C

a What does the graph show about

  i the reaction rates with the two acids?

_____

_____

  ii the amount of hydrogen collected in the two reactions?

_____

_____

[4 marks]

b The rate of reaction can be worked out from the gradient of **Graph C**.

Graph C

B–A*

  i Draw construction lines on the graph and calculate the gradient of the line.

_____

[3 marks]

  ii What does **interpolation** mean?

_____

_____

[1 mark]

Faster or slower (2)

D–C

1 What products are made during an explosion?

_____ [2 marks]

D–C

2 a The reaction between calcium carbonate and hydrochloric acid is measured by the decrease in mass. Look at the equation, why is there a decrease in mass?

$$CaCO_3 + 2HCl \rightarrow CaCl_2 + H_2O + CO_2$$

_____ [2 marks]

b The graph shows how the rate of reaction between calcium carbonate and dilute hydrochloric acid is measured.

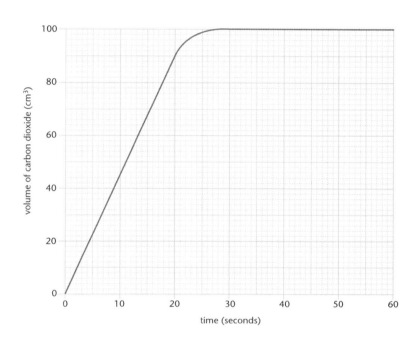

i At which time does the reaction stop? _____ [1 mark]

ii If this reaction were using powdered calcium carbonate, sketch on the graph the line that would show the reaction of the same mass of calcium carbonate as lumps. [2 marks]

iii Why does the reaction slow down? Use ideas about collisions in your answer.

_____ [1 mark]

B–A*

c Why does the reaction rate increase if the surface area increases. Use ideas about collisions between particles in your answer.

_____ [1 mark]

D–C

3 a What does a catalyst do?

_____ [1 mark]

B–A*

b Most catalysts only make a **specific** reaction faster. Explain why.

_____

_____ [2 marks]

C2 Revision checklist

- I know that paint is a colloid where solid particles are dispersed in a liquid, but are not dissolved. ☐

- I know that thermochromic pigments change colour when heated or cooled. ☐

- I know that brick is made from clay, glass from sand, and aluminium and iron from ores. ☐

- I know that the decomposition of limestone is:
  calcium carbonate → calcium oxide + carbon dioxide. ☐

- I know that the outer layer of the Earth is made up of continental plates with oceanic plates under oceans. ☐

- I know that the size of crystals in an igneous rock is related to the rate of cooling of molten rock. ☐

- I know that copper can be extracted by heating its ore with carbon, but purified by electrolysis. ☐

- I know that alloys are mixtures of metals, for example, copper and zinc make brass. ☐

- I know that aluminium does not corrode when wet as it has a protective layer of aluminium oxide. ☐

- I know that iron rusts in air and water to make hydrated iron(III) oxide. ☐

- I know that toxic carbon monoxide comes from incomplete combustion of petrol or diesel in cars. ☐

- I know that catalytic converters can remove CO by conversion:
  $2CO + 2NO \rightarrow N_2 + CO_2$. ☐

- I know that a temperature increase makes particles move faster, so increasing the rate of reaction. ☐

- I know that a catalyst is a substance which changes the rate of reaction but is unchanged at the end. ☐

Collecting energy from the Sun

1 a Write down **four** advantages of using photocells.

_____

_____

_____

_____ [4 marks]

b A p-n junction is made from two pieces of silicon. Explain how the two pieces of silicon are different and what causes the difference.

_____

_____

_____ [3 marks]

2 a During the day, energy from the Sun passes through the large window and warms the room; this is called passive solar heating.
How is the room heated during the night?

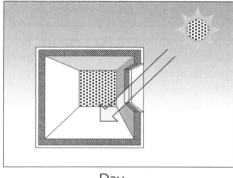

Day

_____ [1 mark]

b The diagram represents the electromagnetic spectrum.

X-rays	ultraviolet	visible light	infrared	radio

i Write the letter **S** on the diagram to show the wavelength of radiation from the Sun that is absorbed by plants in a greenhouse.

ii Write the letter **P** on the diagram to show the wavelength of radiation that is re-radiated from the plants in a greenhouse. [2 marks]

3 a A wind turbine transfers the kinetic energy of the wind into electricity.
What is wind? [1 mark]

_____

b Write down **two** advantages and **two** disadvantages of generating electricity using wind turbines.
Advantages _____

_____

Disadvantages _____

_____ [4 marks]

Generating electricity

1 a The diagram shows a model dynamo. When the coil is spun, a current is produced. Write down **two** ways in which the size of the current can be increased.

_____

_____ [2 marks]

b A model generator consists of a coil of wire rotating between the poles of a magnet. How is the structure of a generator at a power station different from the model generator?

_____ [1 mark]

D–C

2 The flow diagram represents how fossil fuels at a power station provide electrical energy for distribution around the country.

| fossil fuel burned | water heated to produce steam | steam turns turbine | A | generator produces electricity | electricity distributed |

a One step in the process has been missed out. What should be in box **A**?

_____ [1 mark]

b High voltage transmission lines distribute electricity around the country at 400 kV. We use electricity in our homes at 230 V. Explain why electricity is **not** distributed at 230 V.

_____

_____

_____ [3 marks]

c A power station is about 30% efficient. What are the main causes of inefficient energy transfer in a power station?

_____

_____

_____ [3 marks]

D–C

B–A*

Fuels for power

D–C

1 a A nuclear power station uses uranium as its energy source. Uranium is not a fuel.

   i What is a fuel?

   _____ [2 marks]

   ii How does uranium provide energy in the form of heat?

   _____ [1 mark]

B–A*

   b Biomass is an energy source that produces carbon dioxide as a waste product.

   i Suggest a problem that is caused by carbon dioxide.

   _____ [1 mark]

   ii Write down the advantage of using biomass as an energy source.

   _____ [1 mark]

D–C

2 a Each of the headlamp bulbs in Anna's car is connected to a 12 V battery.
When she switches on the headlamps, a current of 2 A passes through the bulb.
Calculate the power rating of the bulb.

   _____ [4 marks]

   b In her home, Anna uses a 2.5 kW kettle for $\frac{1}{2}$ hour each day. Electricity costs
12p per kWh. How much does it cost Anna to use her kettle each day?

   _____ [4 marks]

B–A*

   c Anna sets her dishwasher to work overnight. Why is electricity cheaper during
the night?

   _____ [1 mark]

D–C

3 a Finish the sentences by choosing the **best** words from this list.

   calcium   cancer   DNA   kinetic   nuclear   mutate   react

   Radiation from _____ energy sources causes ionisation.

   This causes a change in the structure of atoms. One of the chemicals in body

   cells is _____ and when this is exposed to radiation it can

   _____ . As a result, body cells may divide in an uncontrolled

   way. This can lead to _____ .   [4 marks]

B–A*

   b Write down the advantages and disadvantages of using uranium as an energy source.

   _____

   _____

   _____ [4 marks]

Nuclear radiations

1 a Background radiation is always present. Write down **two naturally occurring** sources of background radiation.

_____

_____ [2 marks]

b Answer **true** or **false** to each of the following statements about alpha, beta and gamma radiation.

Gamma radiation causes more ionisation than alpha radiation.	
Alpha radiation has a range of a few centimetres in air.	
Beta radiation comes from the nucleus of an atom.	
Beta radiation can be absorbed by a thin sheet of paper.	

[4 marks]

c Atoms are neutral because they contain the same number of positive protons and negative electrons. Explain how negative and positive ions are formed when an atom is exposed to radiation.

Negative ions _____

_____

Positive ions _____

_____ [2 marks]

2 Gamma radiation is used to sterilise medical instruments. It also has other medical uses.

a Write down **one other** medical use for gamma radiation.

_____ [1 mark]

b What property of gamma radiation makes it suitable?

_____ [1 mark]

3 Radioactive waste must be stored securely for possibly thousands of years.

a Why must it be stored for so long?

_____ [1 mark]

b Some people are worried that terrorists may make a nuclear bomb from nuclear waste. Discuss whether or not we should have any concerns about terrorists obtaining nuclear waste.

_____

_____

_____

_____

_____ [3 marks]

Our magnetic field

D–C

1 a Which statement best describes the magnetic field due to a coil of wire?
Put a tick (✓) in the box next to the correct answer.

The magnetic field is circular.	
The magnetic field is radial like the spokes of a bicycle wheel.	
The magnetic field is similar to the field due to a bar magnet.	

[1 mark]

B–A*

b Earth's outer core is mainly molten iron but it is too hot to be magnetic.
Suggest how Earth's magnetic field is produced.

_____

_____

_____

[3 marks]

D–C

2 a Scientists believe that the Moon and Earth used to rotate much faster than they
do now. What do they think caused the rotation to slow down?

_____

[1 mark]

B–A*

b The Moon and Earth were probably formed at the same time. Explain how
scientists believe the Moon and Earth were formed.

_____

_____

_____

_____

_____

[3 marks]

B–A*

3 Cosmic rays originate from the Sun. What happens when they interact with Earth's
magnetic field?

_____

_____

[2 marks]

4 Finish the sentences by choosing the **best** words from this list.

charged particles   **gamma rays**   **hydrogen atoms**   **radio signals**

D–C

Solar flares emit _____ that produce

magnetic fields. The magnetic fields interact with Earth's magnetic field.

B–A*

This causes interference with _____ .

[2 marks]

Exploring our Solar System

1 a On August 24 2006, the International Astronomical Union considered a proposal to redefine planets. There would be twelve planets in our Solar System. Ceres, the largest asteroid would become a planet. Charon, at the moment known as Pluto's moon, would become a 'twin planet' with Pluto. Recently another planet has been discovered beyond Pluto. This proposal was rejected and the decision made that Pluto should no longer be called a planet.

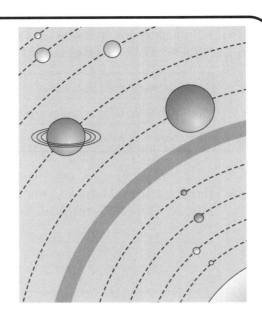

The diagram shows the Sun, the eight planets, Pluto and the asteroid belt.

  i Write the letter **C** on the diagram to show where Ceres orbits.

  ii Write the letter **P** on the diagram to show Pluto.

[2 marks]

b The diagram represents the Moon in orbit around Earth.

  i What is the name of the force that keeps the Moon in orbit around Earth?

_____

_____

[1 mark]

  ii Add an arrow to the diagram to show the direction in which this force acts on the Moon.

_____

[1 mark]

---

2 a Helen Sharman was the first British astronaut. She spent eight days in the Mir Space Station doing science experiments. When she was in the Space Station she experienced **weightlessness** but was not **weightless**.
Why was Helen never weightless?

_____ [1 mark]

b When astronauts work outside a spacecraft, they have to wear special helmets with sun visors. Why do the helmets need special visors?

_____ [1 mark]

c NASA is planning to send a manned spacecraft to another planet before 2030. It is expected to cost £400 billion.
Which planet will the spacecraft go to? Explain the reason for your answer.

_____

_____

[2 marks]

---

3 Proxima Centauri is 4.22 **light-years** away from us.
What is meant by the term **light-year**?

_____ [1 mark]

Threats to Earth

1 a Most asteroids orbit the Sun in a belt between two planets. Which two planets?

D–C

_____ and _____

[2 marks]

b What are asteroids? Put a tick (✓) in the box next to the correct answer.

clouds of very dense gas	
cubes of ice	
rocks that have broken away from planets	
rocks left over when the Solar System formed	

[1 mark]

B–A*

c Asteroids have not joined together to form another planet. Explain why.

_____

_____

[2 marks]

2 The diagram shows the orbits of two bodies orbiting the Sun.
One is a planet, the other is a comet.

D–C

a Label the orbit of the comet with the letter **C**.

[1 mark]

B–A*

b Write the letter **X** to show where on the orbit the comet is travelling at its fastest. [1 mark]

D–C

c Why does a comet's tail always point away from the Sun?

_____ [1 mark]

3 Scientists are constantly updating information on the paths of NEOs.

D–C

a Why is it important to constantly monitor the paths of **NEO**s?

_____

_____ [2 marks]

B–A*

b Scientists may want to change the course of a NEO. Explain how this may
be done.

_____

_____

_____ [2 marks]

The Big Bang

1 a The Universe is expanding. Galaxies in the Universe are moving at different speeds. Which galaxies are moving the fastest?

_____ [1 mark]

D–C

b Scientists know how fast galaxies are moving because they measure **red shift**. What is meant by 'red shift'?

_____
_____
_____ [3 marks]

B–A*

c How does red shift provide information about the speed of a galaxy?

_____
_____ [1 mark]

2 a A star starts its life as a swirling cloud of gas and dust. Describe what happens to this cloud to produce a glowing star.

_____
_____
_____
_____
_____ [4 marks]

B–A*

b A star's life is determined by its size.
i What happens to a medium-sized star, like our Sun, at the end of its life?

_____
_____
_____
_____
_____ [4 marks]

D–C

ii What happens to a very large star at the end of its life?

_____
_____
_____
_____
_____
_____ [4 marks]

P2 Revision checklist

- I can explain how energy from the Sun can be used for heating and producing electricity. ☐

- I can describe how a p-n junction produces electricity. ☐

- I can describe how generators produce electricity. ☐

- I can explain why electricity is distributed via the National Grid at high voltages. ☐

- I can explain the advantages and disadvantages of the fuels used in power stations. ☐

- I can calculate power and the cost of using an electrical appliance for a certain time. ☐

- I know why there is background radiation. ☐

- I can list some uses of alpha, beta and gamma sources and relate their use to their properties. ☐

- I can explain how the Earth's magnetic field is produced and its effect on cosmic rays. ☐

- I can describe how the Moon was formed. ☐

- I can explain why the planets stay in the orbits they do around the Sun. ☐

- I can explain how and why we are exploring space through manned and unmanned spacecraft. ☐

- I can describe asteroids and comets and know the importance of constantly checking NEOs. ☐

- I can describe the Big Bang theory and why scientists believe the Universe is still expanding. ☐

Molecules of life

1 Respiration is a chemical reaction that takes place in the cell.
Write down the name of the part of the cell where respiration takes place.

_____ [1 mark]

2 This diagram shows DNA fingerprints of individuals connected to a robbery.

DNA fingerprints from suspects' blood left at the broken window of a jeweller's shop

Peter   Fred   Sue   James

blood from the scene of the crime

   a Who left blood at the crime scene? Explain your answer.

_____

_____

_____

_____

[2 marks]

   b Describe how a DNA fingerprint is made.

_____

_____

_____

[3 marks]

   c The DNA base code codes for amino acids.

     i How many amino acids are coded for in the following section of DNA?
TATATGTAAAAACAA

_____ [1 mark]

     ii Write down the complementary base sequence for this section of DNA.

_____ [1 mark]

3 Look at the graph. It shows the effect of temperature on an enzyme.

   a What is meant by the term **enzyme**?

_____

_____

_____

_____

[2 marks]

   b Describe the pattern shown in the graph.

_____

_____

[2 marks]

   c What is the optimum temperature of this enzyme?

_____ [1 mark]

   d The enzyme controlled reaction stops at 45 °C. Use ideas about enzyme shape to explain this.

_____

_____

_____ [3 marks]

Diffusion

D–C

1 a What is meant by the term **diffusion**?

_____

_____ [2 marks]

B–A*

b Diffusion takes place between cells of the body.
Describe **three** ways to increase the rate of diffusion between cells.

1 _____

2 _____

3 _____ [3 marks]

D–C

2 a Diffusion takes place in the placenta. Substances diffuse from the foetus into the mother's blood. Write down the name of **two** of these substances.

1 _____

2 _____ [2 marks]

b Diffusion of oxygen into the blood takes place in the lungs.
In which part of the lung does oxygen enter the blood?

_____ [1 mark]

D–C

3 Describe how carbon dioxide moves into the leaf. Use the words **concentration**, **diffusion** and **photosynthesis** in your answer.

_____

_____

_____ [3 marks]

B–A*

4 Look at the picture of a synapse. Describe how the synapse is adapted to carry the signal from one neurone to the next.

_____

_____

_____

_____

_____

_____

[3 marks]

Keep it moving

---

1 Red blood cells are adapted to carry out their function. They are disc-shaped and do not have a nucleus.

a Explain how these adaptations allow them to support their function.

Disc-shaped _____

_____

No nucleus _____

_____ [2 marks]

b Write down the name of the chemical that makes red blood cells red.

_____ [1 mark]

---

2 a Look at the diagram of the heart.

i Draw an **X** to show the part of the heart that receives blood from the lungs.

ii Label the bicuspid valve. [2 marks]

RIGHT   LEFT

b The left ventricle has a thicker wall than the right ventricle. Explain why.

_____

_____ [2 marks]

c People with heart disease may need a heart transplant.

i Describe **two** problems with heart transplants.

1 _____

2 _____ [2 marks]

ii Describe **one** advantage and **one** disadvantage that heart transplants have over heart pacemakers.

Advantage _____

Disadvantage _____ [2 marks]

---

3 a Describe the role of blood vessels in circulating blood around the body.

_____

_____

_____ [3 marks]

b Humans have a double circulatory system. Explain the advantage of a double circulatory system.

_____

_____ [2 marks]

Divide and rule

1 a Amoebas are unicellular organisms. They only have one cell. Humans are multi-cellular; they are made of many cells. It may be a disadvantage to be unicellular rather than multi-cellular. Explain why.

_____

_____ [2 marks]

b The table shows the surface area and volume of different cubes.

i Finish the table by calculating the surface area to volume ratio of each cube. The first one has been done for you.

cube	surface area in cm²	volume in cm³	ratio
A	24	8	24/8 = 3
B	54	27	
C	96	64	
D	150	125	

[3 marks]

ii Organisms made of a single large cell have a disadvantage. Use the information in the table to explain why.

_____

_____ [2 marks]

2 a Write down the name of the type of cell division that makes new **body** cells.

_____ [1 mark]

b Which **two** of the following statements relate to the cell division that makes human body cells? Put ticks (✓) in the **two** correct boxes.

the new cells are diploid ☐

four new cells are made ☐

the new cells contain 23 chromosomes ☐

pairs of chromosomes separate to opposite poles of the cell ☐

the new cell shows variation ☐

chromosomes separate to opposite poles of the cell ☐ [2 marks]

3 a Sperm cells have a structure called an acrosome. Explain why sperm cells need an acrosome.

_____

_____ [2 marks]

b A special type of cell division makes sperm cells.

i Write down the name of this type of cell division.

_____ [1 mark]

ii Describe one way in which this type of cell division is different from the cell division that makes body cells.

_____ [1 mark]

c Sperm cells are haploid. Explain is meant by the term **haploid**.

_____ [1 mark]

Growing up

1 a Both animal and plant cells have a nucleus which makes them similar.
Describe **one other** way that they are similar and **one** way they are different.

Similar _____

_____

Different _____

_____ [2 marks]

b For a fertilised egg to grow into an embryo the cells need to divide
and change.

i Write down the name that is given to cells before they become specialised.

_____ [1 mark]

ii Damaged brain cells cause a disease called Parkinson's. Scientists hope to
repair the damage by taking non-specialised cells from human embryos and
then turning them into brain cells.
Explain why some people may object to this process.

_____

_____ [2 marks]

---

2 a Look at the table. It shows the change in weight of a baby from 0 to 30 months.

age in months	0	3	6	9	12	15	18	21	24	27	30
weight in kg	2.5	5.0	6.4	7.5	8.8	9.6	9.8	10.0	10.1	10.4	10.7

i Plot the points on the graph. The first two have been done for you.

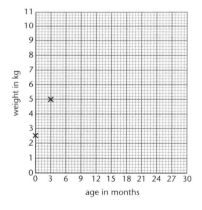

[3 marks]

ii Finish the graph by drawing the best curve. [1 mark]

iii Describe the pattern in the graph.

_____

_____ [2 marks]

iv Write down the phase of human growth shown in the graph.

_____ [1 mark]

b A baby should weigh about 14 kg at 30 months.
Suggest a reason why the baby represented in the graph is underweight.

_____ [1 mark]

Controlling plant growth

D–C

1 a Ben grows apple trees. He decides to take some shoot cuttings from his apple trees and uses rooting powder to grow new trees.
What effect does rooting powder have on the shoot cuttings?

_____ [1 mark]

b Ben also grows wheat and sprays the wheat with selective weedkiller.
The weedkiller destroys the weeds with broad leaves but not the crops.
i How does the weedkiller destroy the weeds?

_____

_____ [2 marks]

ii Explain why the crops are not affected by the weedkiller.

_____

_____ [2 marks]

2 a

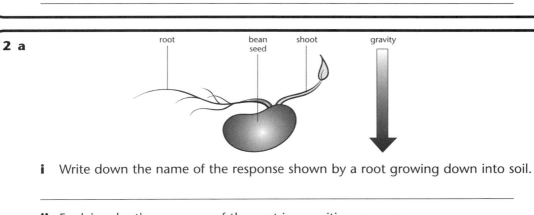

root   bean   shoot   gravity
            seed

i Write down the name of the response shown by a root growing down into soil.

_____ [1 mark]

ii Explain why the response of the root is a positive response.

_____ [1 mark]

D–C

B–A*

b Look at the experiment shown in the diagram.

shoot with tip removed   A   B   light from this side only

seedlings at start   after 2 days

i Write down the name of the hormone that caused the growth of the **shoot**.

_____ [1 mark]

ii How does the hormone cause **shoot B** to bend?

_____

_____

_____ [3 marks]

iii Explain why **shoot A** did not bend.

_____

_____ [2 marks]

New genes for old

1 Richard uses selective breeding to produce apples that are resistant to disease and green in colour. Describe the process of selective breeding.

_____

_____

_____ [3 marks]

D–C

2 When a gene mutates the base sequence changes. Look at the original base code and its mutation.

original base code   **AAAGGTCACTTGAAA**

mutation   **AAAGGTCACTGTAAA**

a Describe the mutation in this base code.

_____ [1 mark]

b Look at the flow chart. It shows how a white pigment is turned into a purple pigment using two enzymes.

```
                    enzyme A                    enzyme B
white pigment ──────────────▶ red pigment ──────────────▶ purple pigment
```

The reactions take place in the petals of a flower. A mutation to the DNA resulted in red flowers. Use ideas about **base sequence**, **proteins** and **enzymes** to explain this.

_____

_____

_____ [3 marks]

B–A*

3 Beta-carotene is needed to produce vitamin A. Beta-carotene is found in carrots but not in rice.

a Describe how genetic engineering can be used to grow rice that provides beta-carotene.

_____

_____

_____ [3 marks]

b Suggest **one** advantage and **one** disadvantage of genetic engineering.

Advantage _____

_____

Disadvantage _____

_____ [2 marks]

D–C

More of the same

1 a Describe how cows are cloned using **embryo transplants**.

_____

_____

_____ [3 marks]

b Scientists are hoping to solve organ transplant problems by cloning pigs.

i Explain how the cloning of pigs could help solve organ transplant problems.

_____

_____ [2 marks]

ii Suggest **one** reason why people may object to this process.

_____ [1 mark]

c Look at the diagram. It shows how Dolly the sheep was cloned.

i Describe **two** ways in which nuclear transfer is different from embryo transplants.

1 _____

_____

_____

2 _____

_____

_____

egg cell taken from sheep A and nucleus removed

cells taken from the udder of sheep B and the nucleus removed

nucleus from sheep B is put into egg of sheep A

egg cell is put into a female sheep to grow

cell grows into a clone of sheep B

[2 marks]

ii The lamb produced is a clone of which sheep, A or B? Explain your answer.

_____ [1 mark]

2 Tissue culture can be used to clone plants.

a Suggest **one** advantage and **one** disadvantage of cloning plants.

Advantage _____

_____

Disadvantage _____

_____ [2 marks]

b Cloning plants is easier than cloning animals. Explain why.

_____

_____

_____ [3 marks]

B3 Revision checklist

- I can interpret data on DNA fingerprinting for identification. ☐

- I can describe DNA replication. ☐

- I know that food and oxygen diffuse across the placenta. ☐

- I can describe diffusion as the net movement of particles from a region of high to low concentration. ☐

- I know that arteries transport blood away from the heart. ☐

- I know that a patient can reject a heart transplant. ☐

- I know that at fertilisation haploid gametes join to form a diploid zygote. ☐

- I know that body cells are made by mitosis and gametes are made by meiosis. ☐

- I can identify the main stages of human growth. ☐

- I know that shoots are positively phototropic and negatively geotropic; roots are the opposite. ☐

- I can describe the stages in selective breeding. ☐

- I know that genetic engineering is used to make insulin. ☐

- I can describe some advantages and disadvantages of cloned plants. ☐

- I know that cloned animals could be used to produce organs for transplants. ☐

What are atoms like?

D–C

1 a What are the particles in the nucleus of an atom?

_____ [1 mark]

b Finish the table to show the relative mass and charge of the atomic particles.

	relative charge	relative mass
electron		0.0005 (zero)
proton	+1	
neutron		

[4 marks]

c What is the atomic number of an atom?

_____ [1 mark]

d What is the mass number?

_____ [1 mark]

B–A*

e What is the atom that has an atomic number of 9, and a mass number of 19, and a neutral charge?

atomic number	9
mass number	19
charge	0
name	

f If an element has the symbol $^{35}_{17}Cl$ how many protons are there in the atom? How many neutrons?

i Protons _____ [1 mark]

ii Neutrons _____ [1 mark]

D–C

2 a What is an isotope?

_____ [1 mark]

B–A*

b Isotopes of an element have different numbers of neutrons in their atoms. Finish the table.

isotope	electrons	protons	neutrons
$^{12}_{6}C$	6	6	
$^{14}_{6}C$	6	6	

[2 marks]

B–A*

3 Draw the electronic structure for the element aluminium. Explain why **three** shells are needed for electrons.

_____

_____

_____ [5 marks]

Ionic bonding

1 a Put a tick (✓) in the box next to the sentence that describes a **metal** atom.

    i An **atom** that has extra electrons in its outer shell and needs to **lose** them to be stable. ☐

    ii An **atom** that has 'spaces' in its outer shell and needs to **gain** electrons to be stable. ☐

    [1 mark]

b Draw a diagram to show how the electrons transfer between a metal atom and a non-metal atom to form a stable pair. **Outer shells only**.

    ◯   ◯

    [3 marks]

c Finish the sentences.

    i If an atom loses electrons a _____ **ion** is formed. [1 mark]

    ii An example of an atom which loses 2 electrons is _____. [1 mark]

d Finish the sentences.

    i A **negative ion** is formed by an atom _____ electrons. [1 mark]

    ii An example of an atom gaining 1 electron is _____. [1 mark]

e Finish the sentences.

During **ionic bonding**, the metal atom becomes a _____

ion and the non-metal atom becomes a _____ ion.

The positive ion and the negative ion then attract one another. They attract to

a number of other ions to make a solid _____. [3 marks]

f Draw the **'dot and cross' model** for sodium chloride.
Outer shell electrons only.

    [4 marks]

g Draw the dot and cross diagram for magnesium chloride.

    [4 marks]

h Put a tick (✓) in the boxes next to the substances that conduct electricity.

    sodium chloride solution ☐   solid sodium chloride ☐

    molten (melted) magnesium oxide ☐   solid magnesium oxide ☐

    molten sodium chloride. ☐   [3 marks]

i Finish the sentences for some of the physical properties of sodium chloride and magnesium oxide.

    i There is a strong attraction between positive and negative ions so they have _____ melting points.

    ii Its ions cannot move in the solid so it does not _____.

    iii Solutions or molten liquids conduct electricity because _____. [3 marks]

D–C

B–A*

D–C

B–A*

Covalent bonding

D–C

1 a Non-metals combine together by **sharing** electrons. What is this type of bonding?

_____ [1 mark]

b Look at the diagram.

Explain how a water molecule is formed from atoms of other elements.

_____
_____

B–A*

_____ [4 marks]

c Carbon dioxide and water do not conduct electricity. Explain why.

_____ [1 mark]

d The formation of simple molecules containing single and double covalent bonds can be represented by dot and cross models.

i Draw the dot and cross diagram of water.

[2 marks]

ii Draw the dot and cross diagram of carbon dioxide.

[2 marks]

[2 marks]

B–A*

2 Carbon dioxide and water have very low melting points. Use the idea of **intermolecular forces** to explain why.

_____
_____ [2 marks]

D–C

3 a Sodium is in group 1. Explain why.

_____ [1 mark]

b Chlorine atoms have 7 electrons in the outer shell. In which group is chlorine?

_____ [1 mark]

c i To which period does fluorine belong?

_____ [1 mark]

ii Explain why.

_____ [1 mark]

B–A*

d The electronic structure of sulphur is 2, 8, 6. In which group is sulphur?

_____ [1 mark]

The group 1 elements

1 a Lithium, sodium and potassium react with water.

    i Which gas is given off?

_____ [1 mark]

    ii They float on the surface. Explain why.

_____ [1 mark]

b Sodium reacts very vigorously with water and forms sodium hydroxide.
Write down the word equation for the reaction of sodium with water.

_____ [2 marks]

c Reactivity of the alkali metals with water increases down group 1.

reactivity increases down		melting point in °C	boiling point in °C
	$_3$Li	179	1317
	$_{11}$Na	98	
	$_{19}$K		774

Estimate the melting point of potassium _____ and

the boiling point of sodium _____. [2 marks]

d Group 1 metals have similar properties. Explain why.

_____ [1 mark]

e Write a balanced symbol equation for the reaction of sodium metal with water.

_____ [2 marks]

---

2 a Marie and Mitch carried out some flame tests. How did they do this?

_____

_____

_____ [4 marks]

b Draw a **straight** line to match up their results.

red   potassium

yellow   lithium

lilac   sodium   [2 marks]

---

3 a Alkali metals have similar properties because when they react their atoms need to lose one electron to form full outer shells. Write down the equation to show the formation of a lithium ion from its atom.

_____ [1 mark]

b Sodium is more reactive than lithium. Explain why. Use ideas about numbers of shells of electrons.

_____

_____ [3 marks]

c What is the process of electron loss called?

_____ [1 mark]

The group 7 elements

1 a There is a **trend** in the **physical appearance** of the halogens at room temperature. Finish the table.

chlorine	
bromine	*orange liquid*
iodine	

[2 marks]

b Group 7 elements have similar properties. Explain why.

_____ [1 mark]

c Chlorine has an electronic structure of 2, 8, 7. It gains one electron to become 2, 8, 8.

i Write an ionic equation to show this.

_____ [1 mark]

ii What is the process of electron gain called?

_____ [1 mark]

iii Fluorine is more reactive than chlorine. Explain why. Use ideas about the gain of electrons.

_____

_____ [2 marks]

2 a When a halogen reacts with an alkali metal a **metal halide** is made. Write down the word equation for the reaction between potassium and iodine.

_____ [2 marks]

b Potassium reacts with chlorine to produce potassium chloride. Write a balanced equation to show this reaction.

_____ [2 marks]

3 a If halogens are bubbled through **solutions of metal halides** there are two possibilities: **no reaction**, or a **displacement reaction**.

i If chlorine is bubbled through potassium bromide solution a red-brown colour is seen. Explain why.

_____ [1 mark]

ii If bromine is bubbled through potassium chloride solution there is no reaction. Explain why.

_____ [1 mark]

b Bromine displaces iodine from potassium iodide solution.

i Write down a word equation for this reaction.

_____ [2 marks]

ii Write down a balanced equation for this reaction.

_____ [2 marks]

Electrolysis

1 a Explain the key features of the electrolysis of dilute sulphuric acid.

_____

_____

_____

_____

_____

_____ [6 marks]

b Explain why the volume of hydrogen gas and the volume of oxygen gas given off in this process are different.

_____ [1 mark]

D–C

2 The electrolysis of sodium chloride takes place in solution.
Describe the reactions at each of the electrodes. Explain the electron transfer processes in each case.

a At the cathode _____

_____

_____ [3 marks]

b At the anode _____

_____

_____ [3 marks]

B–A*

3 a Write about the key features of the production of aluminium by electrolytic decomposition.

_____

_____

_____

_____ [4 marks]

b Write down the word equation for the decomposition of aluminium oxide.

_____ [1 mark]

c The electrode reactions in the electrolytic extraction of aluminium involve the transfer of electrons. Explain how reduction and oxidation occur at the electrodes.

_____

_____

_____

_____ [4 marks]

d Why is the chemical cryolite added to aluminium oxide?

_____

_____ [2 marks]

D–C

B–A*

Transition elements

D–C

1 a A compound that contains a transition element is often coloured.

   i What is the colour of copper compounds?

_____

   ii What is the colour of iron(II) compounds?

_____

   iii What is the colour of iron(III) compounds?

_____ [3 marks]

b A transition metal and its compounds are often catalysts.

   i Which transition metal is used in the Haber process to produce ammonia?

_____ [1 mark]

   ii If the metal used to harden margarine is number 28, suggest whether this is a transition metal or not. Use the periodic table on page 112 to help you.

_____ [1 mark]

c If a transition metal carbonate is heated, it decomposes to form a metal oxide and carbon dioxide. Write down a word equation for the decomposition of copper carbonate.

_____ [1 mark]

D–C

2 a Sodium hydroxide solution is used to identify the presence of transition metal ions in solution. Finish the table.

ion	colour
Cu^{2+}	
Fe^{2+}	
Fe^{3+}	

[3 marks]

B–A*

3 a Copper carbonate decomposes into copper oxide and carbon dioxide. Write the balanced symbol equation for this thermal decomposition.

_____ [2 marks]

b Iron(III) ions react with hydroxide ions to from a precipitate. Write a balanced ionic symbol equation to show this precipitation reaction.

_____ [3 marks]

Metal structure and properties

1 a Silver is often chosen to make a piece of jewellery. Put a (ring) around the **two** properties that are important for this.

ductile   good electrical conductor   high boiling point

high melting point   lustrous   malleable

good thermal conductivity   [2 marks]

b Copper is often used for the base or the whole of a saucepan. Use your knowledge about chemical properties to explain why.

_____

_____

_____ [3 marks]

c Aluminium is used in the aircraft industry and in modern cars. Explain why.

_____ [1 mark]

d Label the diagram to explain a **metallic bond**.

[1 mark]

e Metals have high melting points and boiling points. Use ideas about delocalised electrons to explain why.

_____

_____ [3 marks]

2 a What are superconductors?

_____ [1 mark]

b What are **two** potential benefits of superconductors?

_____

_____ [2 marks]

c Why do scientists need to develop superconductors that will work at 20 °C?

_____ [1 mark]

d Explain why metals conduct electricity.

_____ [1 mark]

e

metal	electrical conductivity 10^6/cm Ω	thermal conductivity W m^{-1} K^{-1}	tensile strength MPa
A	0.099	80	350
B	0.452	320	170
C	0.596	400	210

i Which metal would be used for electrical cables? _____ [1 mark]

ii Which metal would be used for pulling up elevators (lifts)? _____ [1 mark]

C3 Revision checklist

- I know that the nucleus is made up of protons and neutrons, with each having a relative mass of 1. ☐

- I know that electrons surround the nucleus and occupy shells in order. They have almost no mass, 0. ☐

- I know that positive ions are formed by the loss of electrons from the outer shell. ☐

- I know that negative ions are formed by the gain of electrons into the outer shell. ☐

- I know that non-metals combine by sharing electrons, which is called covalent bonding. ☐

- I can draw the 'dot and cross' diagrams of simple molecules such as H_2, Cl_2, CO_2 and H_2O. ☐

- I know that lithium, sodium and potassium react vigorously with water and give off hydrogen. ☐

- I know that group 1 metals have one electron in their outer shell, which is why they are similar. ☐

- I know that chlorine is a green gas, bromine is an orange liquid and iodine is a grey solid. ☐

- I know that halogens gain one electron to form a stable outer shell. This is called reduction. ☐

- I know that in the electrolysis of dilute sulphuric acid, H_2 is made at the cathode and O_2 at the anode. ☐

- I know that when aluminium oxide is electrolysed, Al is formed at the cathode and O_2 at the anode. ☐

- I know that compounds of copper are blue, iron(II) are light green and iron(III) are orange/brown. ☐

- I know that metals conduct electricity as the 'sea' of electrons move through the positive metal ions. ☐

Speed

1 Freddie is on holiday in France. He travels 390 km in 3 hours on a French motorway.

   a Calculate the average speed of his car on this journey in km/h.

_____

_____ [3 marks]

   b Why is your answer the **average** speed of the car?

_____ [1 mark]

   c The speed limit on the motorway is 130 km/h. Did the car break the speed limit? Explain your answer.

_____

_____ [2 marks]

2 a A cycle track is 500 m long. Imran completes 10 laps (he rides round the track 10 times). Imran travelled at an average speed of 12.5 m/s.

    i How long did he take to complete ten laps?

_____

_____ [5 marks]

    ii Imran put on a spurt in the last lap, completing it in 35 s. What was his average speed, in m/s, for the last lap?

_____

_____ [3 marks]

    iii Calculate Imran's average speed for the first nine laps.

_____

_____

_____ [4 marks]

   b The graph shows Ashna's walk to her local shop and home again.

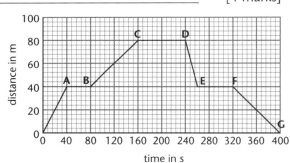

    i During which part of her journey did she walk fastest?

_____ [1 mark]

    ii Calculate Ashna's speed between 0 and A.

_____

_____ [3 marks]

    iii Calculate Ashna's speed between F and G.

_____

_____ [2 marks]

Changing speed

1 Darren is riding his bicycle along a road. The speed-time graph shows how his speed changed during the first minute of his journey.

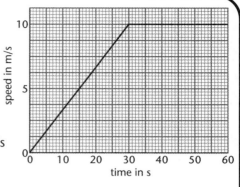

a Darren makes the same journey the next day but:
 – increases his speed at a steady rate for the first 20 s, reaching a speed of 10 m/s
 – travels at a constant speed for 10 s
 – slows down at a steady rate for 15 s to a speed of 5 m/s
 – travels at a constant speed of 5 m/s.
 Plot the graph of this journey on the axes given.

[4 marks]

b How could you calculate the distance Darren travelled in the first minute of his original journey?

_____ [1 mark]

c The way in which the speed of a car changes over a 60 s period is shown in the table.

i Plot a speed-time graph for the car using the axes given. [5 marks]

time in s	speed in m/s
0	0
5	5
10	10
15	15
20	15
25	15
30	15
35	15
40	15
45	11
50	7.5
55	3.5
60	0

ii The car is in an area where the speed limit is 50 km/h. Does the driver exceed this limit? Show clearly how you decided.

_____

_____ [4 marks]

iii Use the graph to calculate the acceleration of the car.

_____

_____ [3 marks]

iv Use the graph to calculate the distance travelled by the car.

_____

_____ [4 marks]

2 a A car accelerates from 10 to 40 m/s in 6 s. Calculate its acceleration.

_____

_____ [4 marks]

b A cheetah accelerates at 6 m/s². How fast is it moving after 5 seconds, starting from rest?

_____

_____ [4 marks]

Forces and motion

1 A car of mass 500 kg accelerates steadily from 0 to 40 m/s in 20 s.

a What is its acceleration?

_____

_____

_____ [4 marks]

D–C

b What resultant force produces this acceleration?

_____

_____ [3 marks]

c The actual force required will be greater than your answer to **b**. Why?

_____ [1 mark]

d Calculate the acceleration of the car if the resultant force is increased to 1250 N.

_____

_____ [3 marks]

B–A*

e The Moon orbits Earth due to gravitational attraction.

i Add an arrow to Earth, labelled F$_E$, to show the force on Earth due to the Moon. [1 mark]

B–A*

ii Add an arrow to the Moon, labelled F$_M$, to show the force on the Moon due to Earth.

Moon [1 mark]

Earth

iii Write down **three** things about this pair of forces.

_____

_____

_____ [3 marks]

2 Helen is driving her car on a busy road when the car in front brakes suddenly. She puts her foot firmly on her brake pedal and just manages to stop without hitting the car in front.

a Write down **two** things, apart from speed, that could increase Helen's thinking distance.

_____ [2 marks]

D–C

b Later that day Helen is driving at high speed on a motorway.

i How will Helen's thinking distance change?

_____ [1 mark]

ii Explain why.

_____ [2 marks]

c Why do worn tyres increase the braking distance?

_____

_____ [2 marks]

B–A*

Work and power

1 a Calculate the amount of work Hilary does when she lifts a parcel of weight 80 N onto a shelf 2 m above the ground.

_____

_____ [3 marks]

b How high could Hilary lift a parcel of weight 60 N if she did the same amount of work?

_____

_____ [3 marks]

c The brakes of a car produce a force of 6000 N.

i Calculate the braking distance if the car has to lose 300 000 J of kinetic energy.

_____

_____

_____ [4 marks]

ii What happens to the kinetic energy lost by the car in braking?

_____ [2 marks]

2 Chris and Abi both have a mass of 60 kg. They both run up a flight of stairs 3 m high. Chris takes 8 s and Abi takes 12 s.

a Calculate Chris' weight. [Take g = 10 N/kg.]

_____ [2 marks]

b How much work does Chris do in running up the stairs?

_____

_____ [3 marks]

c Calculate Chris' power.

_____

_____ [3 marks]

d Priya produces the same power as Abi but she runs up the stairs in 10 s. Calculate Priya's mass.

_____

_____

_____

_____ [5 marks]

3 Why should we keep our fuel consumption to a minimum to protect the environment?

_____

_____ [3 marks]

Energy on the move

1 Use the data about fuel consumption to answer the following questions.

car	fuel	engine size in litres	miles per gallon	
			urban	non-urban
Renault Mégane	petrol	2.0	25	32
Land Rover	petrol	4.2	14	24

a Why is the fuel consumption better in non-urban conditions?

_____

_____ [2 marks]

b How many gallons of petrol would a Land Rover use on a non-urban journey of 96 miles?

_____ [2 marks]

c Would a Renault Mégane use more or less fuel for the same journey? _____ [1 mark]

d Write down a reason for the difference.

_____ [1 mark]

D–C

2 a Which possesses more kinetic energy?

	mass in kg	speed in m/s		mass in kg	speed in m/s
A	2	5	B	2	7
C	20	2	D	15	2

A or B _____   C or D _____ [2 marks]

b Sam is driving a car of mass 1200 kg at a speed of 20 m/s.

i Calculate the kinetic energy of the car.

_____

_____ [4 marks]

ii When Sam suddenly applies the brakes, the car travels 32 m before it stops. Calculate the braking force.

_____

_____ [4 marks]

iii Sam suggests that the braking distance would be 16 m if he was travelling at 10 m/s. Do you agree? Explain your answer.

_____

_____ [2 marks]

D–C

B–A*

3 a How do battery-powered cars cause pollution?

_____ [2 marks]

b Give **one** advantage of solar-powered cars compared to battery-powered cars.

_____ [1 mark]

c Give **one** disadvantage of solar-powered cars compared to battery-powered cars.

_____ [1 mark]

D–C

Crumple zones

D–C

1 Kevin was involved in an accident on the M1 motorway. Luckily, he was not seriously hurt, but his car was badly damaged.

    a Finish the table to show how each feature helps to reduce Kevin's injuries by absorbing energy.

safety feature	how it works
seatbelt	
crumple zones	
air bag	

[3 marks]

    b **i** What is meant by an 'active safety feature' on a car?

        _____

[2 marks]

        ii What is meant by a 'passive safety feature' on a car?

        _____

[2 marks]

    c Traction control is an active safety feature. Explain how it contributes to safety.

    _____

    _____

[2 marks]

B–A*

    d Compare the effect of active and passive safety features on road safety.

    _____

    _____

[2 marks]

D–C

    e **i** What does **ABS** stand for?

        _____

[1 mark]

        ii Why are ABS brakes safer when a driver has to slam his foot on the brakes to stop quickly?

        _____

        _____

[2 marks]

B–A*

2 To minimise injury the forces acting on the people in a car during a car accident must be as small as possible.

    a Explain why this means safety features must reduce the deceleration of the car on impact.

    _____

    _____

[2 marks]

    b Explain how **one** safety feature is designed to reduce the deceleration of the car on impact.

    _____

    _____

[3 marks]

Falling safely

Grades

1 a Charlie drops a golf ball and a ping pong ball from a height of 30 cm above a table. Both balls hit the table together although their masses are different.
Charlie now drops the two balls from a height of 130 cm above the table.
Explain why the golf ball hit the table before the ping pong ball.

_____

_____ [2 marks]

b Sarah is a sky diver. She has a mass of 60 kg.

What is the value of her acceleration just
after leaving the aircraft?

_____ [1 mark]

c i On the diagram, mark and name the forces acting on Sarah as she falls. [2 marks]

ii What can you say about the size of these forces?

_____ [1 mark]

d Sarah's acceleration decreases as she falls. Explain why.

_____ [2 marks]

e i Eventually she is travelling at a constant speed. What is this speed called?

_____ [1 mark]

ii What can you say about the size of the forces acting on her now?

_____ [1 mark]

f Sarah's brother weighs 1000 N. He also sky dives. Will he travel at a larger or smaller constant speed? Explain your answer.

_____

_____

_____ [4 marks]

g Sarah opens her parachute.

i What happens to each of the forces acting on her now?

_____

_____ [2 marks]

ii Explain how Sarah is able to land safely.

_____

_____ [4 marks]

2 a What can you say about the forces acting on a car when it is travelling at a constant speed?

_____ [1 mark]

b Suggest what may affect the maximum speed of a car.

_____ [2 marks]

c What do car designers do to make the maximum speed as large as possible?

_____ [2 marks]

The energy of theme rides

D–C

1 Kate is bouncing a ball. She drops it from A and it rises to **D** after the first bounce.

a Why is **D** much lower than **A**?

_____

_____

[2 marks]

B–A*

b The ball has a mass of 40 g. **A** is 1.2 m above the ground. Calculate the gravitational potential energy of the ball at **A**.

_____

_____

[4 marks]

c Kate drops the ball from increasing heights. She finds there is a maximum height that the ball bounces to. Explain why.

_____

_____

[3 marks]

D–C

2 The diagram shows a roller coaster. The carriages are pulled up to **B** by a motor and then released.

a At which point, **A**, **B**, **C**, **D** or **E**, do the carriages have the greatest gravitational potential energy?

_____ [1 mark]

b At which point, **A**, **B**, **C**, **D** or **E** do the carriages have the greatest kinetic energy?

_____

[1 mark]

c Describe the main energy change as the carriages move from **B** towards **C**.

_____

[2 marks]

d Why must the height of the next peak at **E** be less than that at **B**?

_____

_____

[2 marks]

e The theme park decides to build a faster roller coaster. Suggest how they could modify the design to achieve this, using your knowledge of energy transfers.

_____

_____

_____

[3 marks]

B–A*

3 a Penny has a mass of 65 kg. What is her weight? [Take g = 10 N/kg.]

_____

[2 marks]

b Penny travels to another planet where she only weighs 400 N. Calculate the gravitational field strength on the planet.

_____

_____

[3 marks]

P3 Revision checklist

- I can state and use the formula: speed = distance ÷ time, including a change of subject. ☐

- I can draw and interpret distance-time graphs and speed-time graphs. ☐

- I can use the formula: acceleration = change in speed ÷ time taken, including a change of subject. ☐

- I can state and use the formula: force = mass x acceleration, including a change of subject. ☐

- I can describe the factors that may affect thinking, braking and stopping distances. ☐

- I can state and use the formula: work done = force x distance, including a change of subject. ☐

- I can state and use the formula: power = work done ÷ time, including a change of subject. ☐

- I can use the formula: KE = $\frac{1}{2}mv^2$ and apply it to braking distances and other examples. ☐

- I can interpret data about fuel consumption and explain the factors that affect these figures. ☐

- I can describe and evaluate the effectiveness of various car safety features. ☐

- I can describe how and use forces to explain why falling objects may reach a terminal speed. ☐

- I can recognise that acceleration in free fall (g) is constant. ☐

- I can interpret a gravity ride (roller coaster) in terms of PE, KE and energy transfer. ☐

- I can state and use the formula: weight = mass x g, including a change of subject. ☐

Who planted that there?

1 Look at the diagram. It shows the inside of a leaf.

   a Label

   i the upper epidermis

   ii a palisade cell

   iii a mesophyll cell

   b Put an **S** to show the
   position of a stoma.   [3 marks]

[1 mark]

2 Finish this sentence about gas exchange.
Carbon dioxide enters the leaf and oxygen leaves by _____ .   [1 mark]

3 a A leaf is adapted for photosynthesis. It has broad leaves. This gives a large surface
area to absorb light. Write down **three other** ways they are adapted for photosynthesis.

   _____

   _____

   _____   [3 marks]

   b The cells of a leaf are also adapted for maximum photosynthesis.
   Explain how the cells are adapted.
      i Palisade cells _____

      _____

      _____   [2 marks]

      ii Mesophyll cells _____

      _____

      _____   [2 marks]

Water, water everywhere

1 a Lauren investigates osmosis. She put some potato chips into pure water and some into salt water. She left them for two hours. The diagram shows her results. Explain why the potato chip went floppy in salt water.

pure water

salt water

hard rigid chips

soft floppy chips

_____

_____ [2 marks]

D–C

b Lauren repeats the investigation with red blood cells. Describe what happens to the red blood cells in pure water.

_____

_____ [2 marks]

c Lauren investigates osmosis further. She collects 'skin' cells from an onion. She then puts them in a drop of pure water on a microscope slide. She looks at the cells using a microscope. The diagram shows one of the cells that Lauren sees under the microscope.

cell wall

cytoplasm

B–A*

i The cell is turgid. Use ideas about osmosis to explain this.

_____

_____ [2 marks]

ii Lauren adds salt water to the slide. In the space below draw the onion cell to show how it changes in salt water.

[1 mark]

iii The cell is no longer turgid. Write down the name given to describe the cell in salt water.

_____ [1 mark]

2 Transpiration is the movement of water through a plant. Transpiration provides water for photosynthesis.

D–C

a Describe **two other** uses for water in the plant.

1 _____

2 _____ [2 marks]

b The plant needs to reduce water loss by transpiration. To do this it closes the stomata at night. Explain how the stomata are closed. Use the words **photosynthesis**, **guard cell** and **osmosis** in your answer.

B–A*

_____

_____

_____ [3 marks]

Transport in plants

D–C

1 Look at the diagram. It shows the arrangement of cells in a cross section of a root.

A   B

  a Write down the names of the **two** types of cells shown in the diagram.

      Cell **A** _____

      Cell **B** _____

                                                                     [2 marks]

  b Describe the function of cell **B**.

      _____

      _____ [2 marks]

B–A*

  c Explain how the structure of **B** helps it carry out its function.

      _____

      _____ [2 marks]

D–C

2 Jasmine investigates transpiration in plants. She sets up two lines using this apparatus.

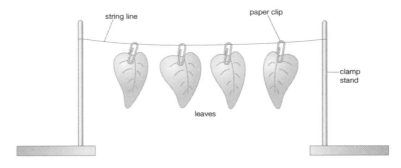

string line   paper clip

clamp stand

leaves

  a One line is kept in the dark and one line is kept in the light. The leaves in the **light** lost 7.4 g in mass. Suggest how much mass the leaves in the **dark** lost.

      _____ [1 mark]

B–A*

  b Jasmine set up a third line and placed a clear plastic bag over the leaves.
      i Describe how this will affect the rate of transpiration.

      _____ [1 mark]

      ii Explain how the plastic bag changed the rate of transpiration.

      _____

      _____ [2 marks]

Plants need minerals too

1 a Finish the table to show why a plant needs certain minerals.
The first one has been done for you.

mineral	why the mineral is needed	how the mineral is used
nitrate	*for cell growth*	*nitrogen is used to make amino acids*
phosphate		
potassium		
magnesium		

[6 marks]

b Mineral deficiency causes poor plant growth. A plant grown without nitrates will have poor growth and yellow leaves. Describe how a plant will look if it is grown without

i phosphate _____

_____ [2 marks]

ii potassium _____

_____ [2 marks]

iii magnesium _____

_____ [1 mark]

2 Look at the graph. It shows the uptake of minerals by algae living in pond water.

a What is the concentration of chlorine
i in the cells of the algae?

_____

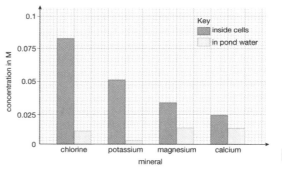

[1 mark]

ii in the pond water?

_____ [1 mark]

b Chlorine is not absorbed by diffusion. Explain why.

_____

_____ [2 marks]

c Name the type of transport that the algae use to take up chlorine from the pond water.

_____ [1 mark]

d Destroying the mitochondria inside the algae stops the uptake of minerals. Explain why.

_____

_____ [2 marks]

Energy flow

D–C

1 Look at the diagram. It shows a pyramid of numbers.

10s of foxes
100s of shrews
1000s of caterpillars
oak tree

a Explain what information is shown in a pyramid of numbers.

_____

_____

_____

[2 marks]

b In the space below draw a pyramid of biomass for the same food chain.

[1 mark]

c As energy flows through the food chain some is lost. Write down **one** way in which energy is lost.

_____

[1 mark]

B–A*

2 Farmers who grow crops produce large amounts of biomass. The biomass has many uses. They may choose to eat the biomass or feed it to their cows.

a Suggest **two other** ways they may choose to use the biomass.

1 _____

2 _____ [2 marks]

b Look at the diagram. It shows the energy transfer from the farmer's crops to the cow.

1022 kJ in heat loss

Sun

energy for growth

1909 kJ in waste

3090 kJ energy in 1 m² of crops

i Calculate the amount of energy used for growth.

Energy used for growth _____kJ   [2 marks]

ii Calculate the efficiency of energy transfer of the cow.

Energy efficiency _____%   [2 marks]

c Some people think it is better to grow crops for food instead of producing meat to eat.
Use your answer to part **b** to explain why.

_____

[1 mark]

D–C

3 a Brazil produces a lot of sugar cane. The sugar cane is used to produce a biofuel. Describe how. Use the words **yeast**, **fermentation** and **petrol** in your answer.

_____

_____

_____

[3 marks]

B–A*

b Describe **one** advantage and **one** disadvantage of using biofuels.

Advantage _____

_____

[1 mark]

Disadvantage _____

_____

[1 mark]

Farming

1 a DDT is a chemical that can be used to kill insects.
Look at the food chain from a marine environment.

plankton ⟶ krill ⟶ penguin ⟶ seal

DDT contaminated the water. The plankton absorbed the DDT but did not die.
The seals began to die from the DDT.
Explain why.

_____

_____

_____ [3 marks]

b Look at the picture. It shows intensive farming of pigs.
This type of farming is more energy efficient than keeping
pigs outside. Explain why.

_____

_____

_____ [2 marks]

D–C

B–A*

2 a Describe how tomatoes can be grown without soil.

_____

_____ [2 marks]

b Write down **one** advantage and **one** disadvantage of growing tomatoes without soil.

Advantage _____

_____

Disadvantage _____

_____ [2 marks]

D–C

B–A*

3 a Organic farmers use a method of farming called crop rotation to help the soil.
Describe **two other** methods of farming used by organic farmers.

1 _____ [1 mark]

2 _____ [1 mark]

b Organic farming is not possible in countries such as Ethiopia.
Suggest why.

_____

_____ [2 marks]

c In 1935 large cane toads were introduced into Australia to control insects feeding on sugar
cane. The toads did not eat the insects; they ate native toads instead. The cane toads have
no predators in Australia.
i Write down the name given to this type of pest control.

_____ [1 mark]

ii The cane toads are now bigger pests than the insects they were sent to control.
Suggest **one** reason why.

_____ [1 mark]

D–C

B–A*

D–C

Decay

1 a Look at the apparatus Shahid uses to investigate decay. Both samples are weighed, left for two days and then re-weighed.

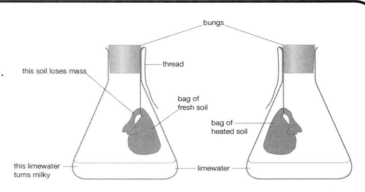

bungs

this soil loses mass

thread

bag of fresh soil

bag of heated soil

this limewater turns milky

limewater

The table shows Shahid's results.

mass in g	bag of fresh soil	bag of heated soil
at start	6.2	6.5
at end	5.8	6.5
change		0

i Calculate the change in mass of the bag of fresh soil. Write your answer in the shaded box. [1 mark]

ii The heated soil did not change in mass. Explain why.

_____

_____ [2 marks]

b Earthworms feed on the remains of dead and decaying organisms. How do earthworms help microorganisms to increase the rate of decay?

_____

_____ [2 marks]

c Shahid extends the investigation. He sets up four more bags of fresh soil and leaves them at different temperatures.

i Predict which of Shahid's bags lost the most mass. Put a (ring) around the correct answer.

bag A 10 °C   **bag B 20 °C**   **bag C 40 °C**   **bag D 60 °C** [1 mark]

ii Explain your answer to part **c i**.

_____

_____ [2 marks]

d Fungi such as mushrooms are called saprophytes. Describe how fungi feed on dead plant material.

_____

_____ [2 marks]

2 Jennifer grows strawberries but she has too many to eat before they decay. She preserves the strawberries by making jam. Explain how this method slows down decay of the strawberries.

_____

_____ [2 marks]

Recycling

1 The diagram shows the carbon cycle.

a Finish labelling the carbon cycle. Choose **three** words from this list.

burning
digestion
feeding
photosynthesis
reproduction
respiration

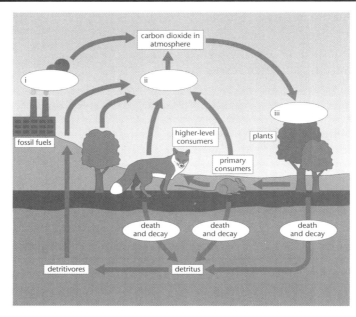

[3 marks]

b Write down the name of the process that removes carbon dioxide from the air.

_____ [1 mark]

c Decomposers return carbon dioxide to the air. Explain why.

_____

_____ [2 marks]

d Look at the picture. It shows marine animals called corals. The carbon in the shells of coral is recycled over millions of years. Explain how this occurs.

_____

_____

_____

_____ [3 marks]

2 In the nitrogen cycle, plants take up nitrates from the soil.

a Why do plants need nitrates?

_____ [1 mark]

b The nitrates are returned to the soil by decomposers. Explain how.

_____

_____ [2 marks]

c Nitrogen gas in the air cannot be used directly by plants. Explain why.

_____ [1 mark]

d Describe **two** ways nitrogen gas can be turned into nitrates for the plants.

1 _____

2 _____ [2 marks]

e Write down the name of the bacteria that return nitrogen to the air.

_____ [1 mark]

B4 Revision checklist

- I can label a diagram showing the parts of a leaf. ☐

- I can explain how a leaf is adapted for photosynthesis. ☐

- I know that osmosis is the movement of water molecules across a partially-permeable membrane. ☐

- I can describe the structure of xylem and phloem. ☐

- I can explain how transpiration rate can be increased. ☐

- I know that active transport needs energy from respiration. ☐

- I can identify mineral deficiencies in plants. ☐

- I can construct pyramids of numbers and biomass. ☐

- I can explain the reasons for developing biofuels. ☐

- I can describe the difference between intensive farming and organic farming. ☐

- I know that a saprophyte decays organisms by releasing enzymes. ☐

- I can explain why the different preservation methods stop food decay. ☐

- I can describe the carbon cycle. ☐

- I can describe the nitrogen cycle. ☐

Acids and bases

1 a i What is an alkali?

_____ [1 mark]

ii Finish the word equation for neutralisation.

_____ + base → salt + _____ [2 marks]

b Write down the word equation for the reaction between copper carbonate and sulphuric acid.

_____ + _____ → _____ + _____ + _____

[3 marks]

c Write down the name of the compound formed when sodium hydroxide reacts with nitric acid.

_____ [1 mark]

d Finish and balance the symbol equations for the reactions between

i $HCl + NaOH \rightarrow$ _____ $+ H_2O$ [1 mark]

ii hydrochloric acid and calcium carbonate $CaCO_3$

_____ [3 marks]

D–C

B–A*

---

2 a Which ion is found in an acid solution?

_____ [1 mark]

b Which ion is found in an alkaline solution?

_____ [1 mark]

c If these **two** types of ion react together, what is made?

_____ [1 mark]

B–A*

---

3 a How does the pH of an acid change when an alkali is added to it?

_____ [1 mark]

b **Universal indicator solution** can be used to measure the acidity of a solution. A few drops are added to the test solution and then the colour of the solution is compared to a standard colour chart.
Describe how the colour changes when a strong acid is added to an alkali to neutralise it.

_____

_____

_____ [3 marks]

D–C

Reacting masses

1 a Work out the relative formula mass of NaOH.

_____ [1 mark]

b Work out the relative formula mass of $CaCO_3$.

_____ [1 mark]

c Work out the relative formula mass of $Ca(OH)_2$.
Use the relative atomic masses.

H 1   C 12   O 16   Na 23   Ca 40

_____ [1 mark]

d Explain why mass is conserved in a chemical reaction.

_____

_____ [2 marks]

2 a Leo and Lesley made some crystals of magnesium sulphate. They did not make as much as they hoped. They wanted to make 42 g. They only made 28 g.
  i What was their 'actual yield'?

_____ [1 mark]

  ii What was their 'predicted yield'?

_____ [1 mark]

  iii How will they calculate their percentage yield?

_____

_____ [1 mark]

  iv What was their percentage yield?

_____

_____ [2 marks]

b Zinc carbonate decomposes on heating to give zinc oxide and carbon dioxide.
  i Write a balanced symbol equation for this reaction.

_____ [2 marks]

  ii Calculate how much CO_2 is made when 12.50 g of $ZnCO_3$ decomposes.

_____

_____

_____

_____ [4 marks]

Fertilisers and crop yield

1 a Why do farmers add fertilisers to their crops?

_____ [1 mark]

D–C

b How do fertilisers get into the plants?

_____ [1 mark]

c Why is nitrogen needed for increased plant growth?

_____ [1 mark]

B–A*

---

2 a To calculate the yield when making a fertiliser you need to calculate its **relative formula mass**. What is the relative formula mass of ammonium sulphate $(NH_4)_2SO_4$? Use the periodic table on page 238.

_____ [2 marks]

D–C

b Farmers can use relative formula masses to find the percentage of each element in a fertiliser – it is printed on the bag for them. They use the formula:

$$\text{percentage of element} = \frac{\text{mass of the element in the formula}}{\text{relative formula mass}} \times 100$$

Calculate the percentage of nitrogen in potassium nitrate, KNO_3.

_____

_____

_____ [3 marks]

B–A*

---

3 a Many fertilisers are **salts**, so they can be made by reacting acids with bases. What else is made?

acid + base → salt + _____ [1 mark]

D–C

b Don and Demi want to make some ammonium phosphate.

i Which acid will they need to use?_____ [1 mark]

ii Which alkali will they need to use? _____ [1 mark]

iii Write down a word equation to show this reaction.

_____ [2 marks]

c Describe in steps how Don and Demi make ammonium phosphate from their acid and alkali.

_____

_____

_____

_____ [6 marks]

B–A*

---

4 Too much fertilisers applied may increase the level of nitrates or phosphates in a water course. This may cause eutrophication to occur. Explain how this happens.

_____

_____

_____ [3 marks]

B–A*

The Haber process

1 a Write about how ammonia is made. Include the conditions needed in your answer.

_____

_____

_____

[3 marks]

b As the reaction for producing ammonia is reversible, the percentage yield for the reaction cannot be 100%.

i How does the pressure affect the reaction?

_____

_____

[2 marks]

ii How does the temperature affect the reaction?

_____

_____

[2 marks]

iii Give the optimum temperature for this reaction and explain why it is used.

_____

_____

_____

[2 marks]

2 a Look at the graph. At which pressure is most ammonia made at 400 °C ?

Percentage of ammonia made

_____ [1 mark]

b As the pressure increases what happens to the yield of ammonia?

_____ [1 mark]

c As the temperature increases what happens to the yield of ammonia?

_____ [1 mark]

d Chemical plants need to work at the conditions that produce the highest percentage yield for a reaction, most cheaply. Use ideas about optimum conditions to explain how this is achieved in the manufacture of ammonia. Explain why these cut costs.

_____

_____

_____

_____

[5 marks]

Detergents

1 a A detergent can be made by **neutralising** an organic acid using an alkali.
Write a word equation for this reaction.

_____

[2 marks]

b Why are detergents used to clean greasy plates?

_____

_____

[2 marks]

c New washing powders allow clothes to be washed at low temperatures.
i This is good for the environment. Explain why.

_____

_____

[3 marks]

ii It is also good for coloured clothes to be washed at low temperatures.
Explain why.

_____

_____

[2 marks]

d When clothes are washed,
grease is lifted off the
clothes and put into water
by detergent.
Use the words **hydrophilic**
and **hydrophobic** to
explain the stages in
the diagrams.

[3 marks]

2 a The forces that hold molecules of grease together and molecules of dry-cleaning solvent
together are forces between molecules. What are these are called?

_____

[1 mark]

b Molecules of water are held together by stronger forces called **hydrogen bonds**.
The water molecules cannot stick to the grease because they are sticking to each
other much too strongly. On the diagram mark the **covalent** bonds between atoms
and the **hydrogen** bonds between water molecules.

water sticks strongly
to water

water sticks to water too
strongly to stick to grease

[5 marks]

Batch or continuous?

1 a Why are pharmaceuticals made in small batches?

_____ [1 mark]

b How is the large scale production of ammonia different from the small scale production of pharmaceuticals?

_____ [1 mark]

c A **continuous** process plant is effective. Explain why.

_____

_____

_____

_____ [4 marks]

d **Batch** processes are not as efficient but have one major advantage. Explain what this is.

_____ [1 mark]

2 Draw **straight** lines to match the reasons for the high costs of making and developing medicine and pharmaceutical drugs to the **best** explanation.

strict safety laws		The medicines are made by a batch process so less automation can be used.
research and development		They may be rare and costly.
raw materials		They take years to develop.
labour intensive		People need to feel a benefit without too many side effects.

[3 marks]

3 Whether or not a drug is developed depends on a number of economic considerations. Explain what these are.

_____

_____

_____

_____

_____ [6 marks]

Nanochemistry

1 a Draw a **straight** line to match the carbon to its correct structure.

diamond

graphite

buckminster
fullerene

[3 marks]

b Finish the table to show **one** use of the carbon type and a reason for that use.

	diamond	graphite	buckminster fullerene
use			*semiconductors in electrical circuits good electrical*
reason			*conductor on small scale*

[4 marks]

c Explain why these structures have these properties.

diamond	graphite
Does not conduct electricity because _____ _____ _____	Slippery because _____ _____ _____

[2 marks]

2 a What are different forms of the same element called?

_____ [1 mark]

b Explain why the giant structure of diamond is so hard and does not conduct electricity.

_____

_____

_____

_____ [5 marks]

3 a Write down **two** uses of nanotubes.

_____

_____ [2 marks]

b Why are nanotubes so useful as catalysts?

_____ [1 mark]

How pure is our water?

D–C

1 a The water in a river is cloudy and often not fit to drink. To make clean drinking water it is passed through a **water purification** works.
Label the **three** main parts of this process.

[_____] → [_____] → [_____]

[3 marks]

b Explain the stages of water purification in detail.

_____

_____

B–A*

_____

_____

_____

[3 marks]

c i Seawater has many substances dissolved in it so it is undrinkable. Which special technique has to be used to remove these unwanted substances?

_____

[1 mark]

ii Why is this technique not used for all purification of water?

_____

_____

[1 mark]

D–C

2 a Explain why relief organisations concentrate on providing clean water supplies.

_____

_____

[2 marks]

b Water is a **renewable resource** but not an endless resource in any one country. Explain why.

_____

_____

[2 marks]

D–C

3 a Write a word equation for the precipitation reaction between lead nitrate and potassium chloride.

_____

[2 marks]

B–A*

b Write a word equation for the precipitation reaction between silver nitrate $AgNO_{3(aq)}$ and potassium bromide $KBr_{(aq)}$.

_____

[1 mark]

c Write a full balanced symbol equation for this precipitation reaction.

_____

[3 marks]

C4 Revision checklist

- I know that neutralisation is a reaction where: acid + base → salt + water. ☐

- I can construct balanced symbol equations, such as:
  $2KOH + H_2SO_4 \rightarrow K_2SO_4 + H_2O$ ☐

- I can work out the relative formula mass of a substance from its formula, such as $Ca(OH)_2$. ☐

- I can work out the percentage yield using the formula:
  % yield = actual yield × 100 ÷ predicted yield. ☐

- I know that fertilisers provide extra essential elements but excess can cause eutrophication. ☐

- I know that ammonia is made by the Haber process where N_2 and H_2 are put over an iron catalyst. ☐

- I know that the higher the pressure, the higher the percentage yield of ammonia. ☐

- I know that a catalyst will increase the rate of reaction but will not change the percentage yield. ☐

- I know that a detergent has a hydrophilic head and a hydrophobic tail. ☐

- I know that dry cleaning is a process used to clean clothes using a solvent that is not water. ☐

- I know that a continuous process makes chemicals all the time but a batch process does not. ☐

- I can recognise the three allotropes of carbon: diamond, graphite and buckminster fullerene. ☐

- I can explain that graphite is slippery and is used as electrodes as it conducts electricity. ☐

- I know that water purification includes filtration, sedimentation and chlorination. ☐

Sparks!

1 a Sally stands on an insulating mat. She puts her hands on the dome of an uncharged Van de Graaff generator. The Van de Graaff generator is switched on and Sally's hair starts to stand on end.

   i Why does Sally stand on an insulating mat?

_____ [1 mark]

   ii Why must the Van de Graaff generator be uncharged when Sally puts her hands on it?

_____ [1 mark]

   iii What happens to Sally when the Van de Graaff generator is switched on?

_____ [1 mark]

   iv Why does this make Sally's hair stand on end?

_____
_____ [2 marks]

b Donna hangs two small, light plastic balls on nylon threads side by side. She touches each ball with a charged polythene strip.

Why do the balls repel each other?

before   after

_____
_____ [2 marks]

c Paul does the same experiment but this time he only touches one ball with a charged polythene strip. The balls move towards each other but do not touch.

Give as much detail as you can to explain why this happens.

_____
_____ [2 marks]

2 a Use your knowledge of electrostatics to explain the following.

   i You sometimes get an electric shock on closing a car door after a journey.

_____
_____ [2 marks]

   ii You should never shelter under a tree during a thunderstorm.

_____ [1 mark]

   iii Cling film often sticks to itself as it is unrolled.

_____ [1 mark]

b A factory uses machinery with moving parts. How can the machinery become charged?

_____ [1 mark]

c Why do the operators stand on rubber mats?

_____
_____ [3 marks]

Uses of electrostatics

1 A defibrillator delivers an electric shock through the chest wall to the heart.

a What does the electric shock do to the heart?_____

[1 mark]

b The paddles of a defibrillator, charged from a high voltage supply, are placed on the patient's chest. How does the operator ensure that there is good electrical contact with the patient's chest?

_____

_____ [2 marks]

c A current of about 50 A passes through the patient for about 4 ms (0.004 s). In general, such a large current would be fatal.
Why can it be used in this situation?

_____ [1 mark]

d A typical defibrillator has a power of about 100 kW. How much energy does the patient receive if the supply is switched on for 4 ms (0.004 s)?

_____

_____ [4 marks]

2 In a bicycle factory the frames are painted using an electrostatic sprayer. The paint is positively charged. The frames are given the opposite charge to the paint.

a Why does the paint spread out on leaving the sprayer?

_____

_____

_____

_____ [2 marks]

nozzle is charged up positively

object to be painted is negatively charged

b Why are the bicycle frames given the opposite charge to the paint?

_____ [2 marks]

c Freddie decides to repaint his bicycle frame. He uses a paint spray that charges the paint positively but he does **not** charge the frame.
i What charge, if any, does Freddie's bicycle frame acquire? _____ [1 mark]

ii What problem does this cause?

_____ [1 mark]

3 An electrostatic precipitator is placed in the chimney of a power station. It contains some wires and plates which are connected to a high voltage supply.

a If the wires are given a negative charge, what charge must be given to the plates?

_____ [1 mark]

b What charge do the soot particles gain as they pass close to the wires?

_____ [1 mark]

Safe electricals

D–C

1 a Meera sets up the circuit shown.

   i What effect will this have on the brightness of the lamp?

   _____ [1 mark]

variable resistor

   ii Add an ammeter and voltmeter to Meera's circuit to allow her to measure the current through the lamp and the potential difference across the lamp. [2 marks]

   iii If the voltmeter reads 6 V and the ammeter 0.25 A, calculate the resistance of the lamp.

   _____

   _____ [4 marks]

B–A*

b Meera increases the potential difference across the lamp to 12 V. The current increases to 0.4 A.

   i What is the resistance now?

   _____

   _____ [2 marks]

   ii Explain the change in resistance in terms of electron movement.

   _____

   _____

   _____ [4 marks]

D–C

2 A battery has positive and negative terminals. Give **two** differences between the voltage from a battery and mains voltage.

   _____

   _____ [2 marks]

D–C

3 a i Label the live (L), neutral (N) and earth (E) wires in the plug shown. [3 marks]

   ii Which wire, live, neutral or earth, is a safety wire?

   _____ [1 mark]

   iii How does it work?

   _____

   _____ [2 marks]

b i An electric kettle passes a current of 10.5 A when working normally. Should the plug contain a 5 A or 13 A fuse? _____ [1 mark]

B–A*

   ii Why are fuses always connected in the live wire?

   _____

   _____ [2 marks]

   iii The kettle is made from metal. How do the fuse and earth wire stop a person receiving an electric shock if they touch the kettle when it is faulty?

   _____

   _____ [4 marks]

Ultrasound

1 a Sound is a longitudinal wave.

   i Explain how sound travels through the air to reach your ear.

   _____

   _____ [2 marks]

   ii How does the frequency of a note change if its pitch increases?

   _____ [1 mark]

b What is 'ultrasound'?

   _____ [1 mark]

c Light is an example of a transverse wave. What is the difference between a longitudinal and a transverse wave?

   _____

   _____ [2 marks]

D–C

B–A*

2 a Finish the sentences about an ultrasound scan. Choose words from this list.

| **echoes** | **gel** | **image** | **probe** | **pulse** |
| **reflected** | **skin** | **tissues** | **ultrasound** | |

A _____ of ultrasound is sent into a patient's body. At each boundary between different

_____ some ultrasound is _____ and the rest is transmitted. The returning _____

are used to build up an _____ of the internal structure. A _____ is placed on the

patient's body between the ultrasound _____ and their _____. Without it nearly all

the _____ would be _____ by the _____ . [6 marks]

b Give **two** factors that affect the amount of ultrasound sent back to the detector at each interface within the body.

   _____ [2 marks]

c Air has a density of 1.3 kg/m^3. Soft tissue has an average density of 1060 kg/m^3.

   i Explain why the gel used should have a density of about 1060 kg/m^3.

   _____

   _____ [2 marks]

   ii The time delay for an echo from ultrasound in soft tissue at a depth of 0.16 m was 0.2 ms (0.0002 s). Calculate the speed of ultrasound in soft tissue.

   _____

   _____ [4 marks]

D–C

B–A*

3 High-powered ultrasound is used to treat a patient with kidney stones.

a How does ultrasound do this?

   _____

   _____ [3 marks]

b Why must **high-powered** ultrasound be used?

   _____ [2 marks]

D–C

Treatment

D–C

1 a Why are X-rays and gamma rays suitable to treat cancer patients?

_____

_____ [2 marks]

b Why are alpha and beta particles **not** suitable to treat cancer patients?

_____

_____ [2 marks]

B–A*

c Explain how gamma radiation is emitted by the nuclei of a radioactive substance.

_____

_____ [3 marks]

D–C

2 a What is a radioactive tracer?

_____

_____ [2 marks]

b Why is it used?

_____

_____ [2 marks]

c What sort of radiation should a tracer emit?

_____ [1 mark]

d Which organ of the body is investigated using iodine-123 as a tracer?

_____ [1 mark]

e X-rays and gamma rays have similar properties. Why are gamma rays suitable to use as tracers but X-rays are not?

_____ [1 mark]

B–A*

3 Three similar sources of radiation are used to destroy a brain tumour.

a i What type of radiation would be most suitable?

_____ [1 mark]

ii Give a reason for your choice.

_____

_____

[2 marks]

source of radiation

tumour

source of radiation

source of radiation

b Why are there **three** sources of radiation arranged as shown?

_____

_____ [2 marks]

c Describe an alternative technique to destroy a brain tumour that uses only **one** source of radiation.

_____

_____ [2 marks]

What is radioactivity?

D–C

1 a Finish the table about the three types of nuclear radiation.

type of radiation	charge (+, – or 0)	what it is	particle or wave
alpha			
beta			
gamma			

[5 marks]

b Name the type of nuclear radiation that

is the most penetrating _____

is stopped by several sheets of paper _____

has the greatest mass _____

does not change the composition of the nucleus _____

travels at about one-tenth the speed of light _____

[5 marks]

B–A*

2 a What is meant by 'half-life'?

_____

_____

[2 marks]

b The graph shows how the activity of cobalt-60 changes with time. Use the graph to find the half-life of cobalt-60. Show clearly, on the graph, how you got your answer. _____

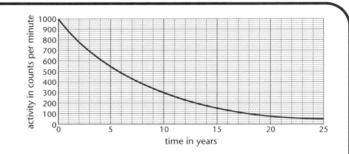

[2 marks]

B–A*

3 Radon, $^{220}_{86}$Rn, is radioactive. It decays to an isotope of polonium, $^{216}_{84}$Po with a half-life of 52 s.

a How many protons are there in a radon nucleus? _____ [1 mark]

b How many neutrons are there in a radon nucleus? _____ [1 mark]

c What is the name of the particle emitted in this decay? _____ [1 mark]

d i Write a nuclear equation to describe the decay of a radon nucleus.

_____ [3 marks]

ii Show that the atomic mass and mass number are conserved in this decay.

_____ [2 marks]

e When a beta particle is emitted from a nucleus the mass number is unchanged but the atomic number increases by one. Explain how this is possible.

_____

_____ [2 marks]

f Finish the nuclear equation for the decay of iodine-131, emitting a beta particle.

$$^{131}_{53}I \longrightarrow {}^{x}_{y}Xe + {}^{0}_{-1}\beta$$

x = _____

y = _____ [2 marks]

Uses of radioisotopes

D–C

1 a Suggest **two natural** sources of background radiation.

_____ [2 marks]

B–A*

b Suggest **two** sources of background radiation that arise from human activity.

_____ [2 marks]

B–A*

2 Harry works for an oil company. A leak has been reported in an underground pipe. He decides to locate the leak by introducing a small amount of radioisotope into the pipe.

a What sort of radiation should the radioisotope emit? Explain your choice.

_____

_____ [3 marks]

b What radiation detector could Harry use?

_____ [1 mark]

c How will Harry tell the site of the leak from his results?

_____ [1 mark]

D–C

3 Smoke alarms use a source of alpha radiation in a small chamber.

a Why is alpha radiation more suitable than either gamma or beta radiation for use in a smoke alarm?

_____ [1 mark]

oppositely charged plates

americium-241 source of alpha particles

b Explain how the smoke alarm works.

_____

_____

_____ [4 marks]

4 Trees contain carbon-14 which is radioactive. The graph shows how the activity of 1 kg of wood changes after a tree has died. This can be used to estimate the age of objects that were once alive.

D–C

a This method cannot be used to date a wooden bowl made from a tree that died less than 200 years ago. Explain why.

_____

_____ [1 mark]

activity in becquerel

time in thousands of years

b The remains of an ancient civilisation were found.
Put a (ring) around the objects that could be dated using this method.

  animal bone   **bronze tool**   **seeds**   **stone**   [2 marks]

c Why can carbon-14 not be used to date rocks?

_____ [1 mark]

B–A*

d 1 kg of wood found in an archaeological dig had an activity of 150 Bq. Use the graph to estimate its age. Show your answer on the graph. [2 marks]

Fission

1 a Use these words to complete the following sentences explaining how a power station works.

generator   source of energy   steam   turbine   water

The _____ provides heat to boil the _____ to produce _____.

The pressure of the _____ turns the _____ which turns

the _____ making electricity.   [6 marks]

b What is meant by 'fission'?

_____
[2 marks]

c Enriched uranium is used as the fuel in a nuclear power station.
What is 'enriched uranium'?

_____
[2 marks]

d A slow neutron can be captured by a uranium-235 nucleus and the 'new' nucleus
then becomes unstable.

i Describe what happens after the capture of the neutron by the uranium nucleus.

_____

_____
[3 marks]

ii How can this one event become a continuous chain reaction?

_____

_____
[2 marks]

2 Nuclear power stations produce radioactive waste.

a Suggest **one** method of disposing of low level radioactive waste.

_____
[1 mark]

b Suggest **one** method of disposing of high level radioactive waste, such as spent
fuel rods from a nuclear reactor.

_____
[1 mark]

3 a In one type of nuclear power station graphite is used as a moderator and boron rods are used
to control the number of nuclear fissions in a given time.

i What does a 'moderator' do?

_____
[1 mark]

ii Why is a moderator necessary?

_____
[1 mark]

iii How do the boron rods control the number of fissions?

_____

_____
[2 marks]

b State **one** advantage and **one** disadvantage of using nuclear energy to generate
electricity compared with using coal as a fuel.

Advantage _____

Disadvantage _____
[2 marks]

P4 Revision checklist

- I can describe static electricity in terms of the movement of electrons. ☐

- I can explain how static electricity can be dangerous and how the risk of shock can be reduced. ☐

- I can describe and explain some uses of static electricity. ☐

- I can explain the behaviour of simple circuits and how resistors are used in circuits. ☐

- I can state and use the formula: resistance = voltage ÷ current, including a change of subject. ☐

- I can explain the function of live, neutral and earth wires, circuit breakers and double insulation. ☐

- I can explain how a wire fuse protects an appliance. ☐

- I can describe the motion of particles in longitudinal and transverse waves. ☐

- I can explain how ultrasound is used in body scans and to break down kidney stones. ☐

- I can explain how nuclear radiation is used in hospitals to treat cancer and as a tracer. ☐

- I can describe radioactive decay and explain and use the concept of half-life. ☐

- I can describe alpha and beta decay and construct nuclear equations to explain it. ☐

- I can explain the origins of background radiation and some non-medical uses of radioisotopes. ☐

- I can describe in detail how domestic electricity is generated in a nuclear power station. ☐

The periodic table

Key

1	relative atomic mass
H	**atomic symbol**
hydrogen	name
1	atomic (proton) number

1	2											3	4	5	6	7	8
																	4 **He** helium 2
7 **Li** lithium 3	9 **Be** beryllium 4											11 **B** boron 5	12 **C** carbon 6	14 **N** nitrogen 7	16 **O** oxygen 8	19 **F** fluorine 9	20 **Ne** neon 10
23 **Na** sodium 11	24 **Mg** magnesium 12											27 **Al** aluminium 13	28 **Si** silicon 14	31 **P** phosphorus 15	32 **S** sulfur 16	35.5 **Cl** chlorine 17	40 **Ar** argon 18
39 **K** potassium 19	40 **Ca** calcium 20	45 **Sc** scandium 21	48 **Ti** titanium 22	51 **V** vanadium 23	52 **Cr** chromium 24	55 **Mn** manganese 25	56 **Fe** iron 26	59 **Co** cobalt 27	59 **Ni** nickel 28	63.5 **Cu** copper 29	65 **Zn** zinc 30	70 **Ga** gallium 31	73 **Ge** germanium 32	75 **As** arsenic 33	79 **Se** selenium 34	80 **Br** bromine 35	84 **Kr** krypton 36
85 **Rb** rubidium 37	88 **Sr** strontium 38	89 **Y** yttrium 39	91 **Zr** zirconium 40	93 **Nb** niobium 41	96 **Mo** molybdenum 42	[98] **Tc** technetium 43	101 **Ru** ruthenium 44	103 **Rh** rhodium 45	106 **Pd** palladium 46	108 **Ag** silver 47	112 **Cd** cadmium 48	115 **In** indium 49	119 **Sn** tin 50	122 **Sb** antimony 51	128 **Te** tellurium 52	127 **I** iodine 53	131 **Xe** xenon 54
133 **Cs** caesium 55	137 **Ba** barium 56	139 **La*** lanthanum 57	178 **Hf** hafnium 72	181 **Ta** tantalum 73	184 **W** tungsten 74	186 **Re** rhenium 75	190 **Os** osmium 76	192 **Ir** iridium 77	195 **Pt** platinum 78	197 **Au** gold 79	201 **Hg** mercury 80	204 **Tl** thallium 81	207 **Pb** lead 82	209 **Bi** bismuth 83	[209] **Po** polonium 84	[210] **At** astatine 85	[222] **Rn** radon 86
[223] **Fr** francium 87	[226] **Ra** radium 88	[227] **Ac*** actinium 89	[261] **Rf** rutherfordium 104	[262] **Db** dubnium 105	[266] **Sg** seaborgium 106	[264] **Bh** bohrium 107	[277] **Hs** hassium 108	[268] **Mt** meitnerium 109	[271] **Ds** darmstadtium 110	[272] **Rg** roentgenium 111							

Elements with atomic numbers 112–116 have been reported but not fully authenticated.

* The Lanthanides (atomic numbers 58–71) and the Actinides (atomic numbers 90–103) have been omitted.

Cu and Cl have not been rounded to the nearest whole number.

Notes

Notes

Workbook answers

B1 Understanding ourselves

Page 130 Fit for life

1 a i Lactic acid
   ii Body needs more energy; not enough oxygen in muscles; anaerobic respiration takes place
   b Takes in extra oxygen to break down the lactic acid; repay oxygen debt
2 Bacteria/viruses can still enter body
3 a Take more exercise; eat a balanced diet; lose weight; drink less alcohol; relax more *(Any 2 = 1 mark each)*
   b Small blood vessels may burst in brain; causing a stroke; blood vessels in kidney may burst *(Any 2 = 1 mark each)*

Page 131 What's for lunch?

1 a Kwashiorkor
   b RDA = 40 x 0.75
      = 30 g
2 a More exercise
   b Meat is a good source of essential amino acids or vitamin B1; he will need to eat alternatives
3 a

food type	enzyme	product
protein	protease	amino acids
fat	lipase	fatty acids and glycerol

(1 mark for each correct box)

   b Bile emulsifies the fat; breaking it up into smaller droplets; lipase enzyme then breaks up the fat molecules; into fatty acids and glycerol *(Any 3)*

Page 132 Keeping healthy

1 a i Vector   ii Feeds off living host
   b Drain stagnate water; to kill larva or put oil on water; to prevent lava from breeding or spay insecticide; to kill adult take larium; to kill protozoan *(Any 1)*
2 a Active: given pathogen; body makes antibodies; long lasting; given antibodies; passive: body does not learn how to make antibodies; short lived
   b Harmless pathogen contains antigen; body learns how to make correct antibody
   c i Do not kill virus; only treat bacterial or fungal infections
      ii Bacteria are becoming resistant
      iii Some patients take the new drug; others take the placebo; without knowing

Page 133 Keeping in touch

1 a

part of the eye	function
retina	contains light receptors
optic nerve	carries impulses to brain
cornea	refracts light

   b Allows it to judge distance
   c Ciliary muscle relaxes; suspensory muscle pulls; lens thin
   d Concave lens; glasses; contact lenses; cornea surgery *(Any 2)*
2 a 3: sensory neurone; 4: relay neurone; 5: motor neurone
   b Impulse triggers release transmitter substance; transmitter substance (acetylrcholine) diffuses across synapse; binds with receptors; triggers impulse in second neurone *(Any 2 = 1 mark each)*

Page 134 Drugs and you

1 a

type of drug	example
hallucinogen	cannabis
depressant	alcohol

(2 correct = 2; 1 correct = 1 mark)

   b For: used as a pain killer; may prevent drug crimes; against: may lead to use of more dangerous drugs

2 a Stops cilia moving; dust collects; leads to smokers' cough
   b Stimulates the receptors; allows more impulses to pass
3 a Matthew; because he drinks 5 units and Jo drinks only 4 units
   b More people drink at weekend; excessive drinking leads to more accidents

Page 135 Staying in balance

1 a Get too hot; start to sweat; lose too much water
   b Hypothermia
   c Blood capillaries in skin dilate; increasing blood flow
   d Body gets too hot mechanisms (e.g. sweating) cool body; body cools switching off mechanisms (stop sweating)
2 a Diabetes
   b Diet/insulin injections
3 a Increases thickness
   b Maintains levels
   c Given sex hormones

Page 136 Gene control

1 a Nucleus; genes; DNA
2 a 10
   b 20
   c Have 23 chromosomes not 10
   d Gene is switched off in nucleus of ear cells
3 a 4
   b 30; because A always links to a T
   c Base sequence would be different; different amino acid coded for; amino acids will form different protein (enzyme); incorrect enzyme so pigment cannot be made *(Any 3 = 1 mark each)*

Page 137 Who am I?

1 a

egg	sperm	fertilised egg	gender of child
X	X	XX	girl
X	Y	XY	boy

(1 mark each)

   b Only males have Y chromosome; only sperm can carry X or Y; eggs can only carry X chromosome *(Any 2 = 1 mark each)*
2 a Radiation; chemicals (e.g. tobacco smoke) *(Any 1)*
   b Base sequence would be different; different amino acid coded for; amino acids will form different protein; incorrect protein so blood cannot clot *(Any 2 = 1 mark each)*
3 a Purple because F1 are all purple
   b i
      ii *(Ring round nn)*

C1 Carbon chemistry

Page 139 Cooking

1 The texture of food is improved; the taste of food is improved; the flavour of food is enhanced; food is easier to digest *(Any 3 = 1 mark each)*
2 a Potatoes; flour
   b Meat; eggs
   c The protein molecules change shape; the shape change is irreversible; the protein molecule is said to be denatured; this changes the appearance/texture of the protein

3 a i Sodium hydrogencarbonate $\xrightarrow{\text{heat}}$ sodium carbonate + carbon dioxide + water
  ii Sodium hydrogencarbonate
  iii Sodium carbonate; carbon dioxide; water
  b $2NaHCO_3 \longrightarrow Na_2CO_3 + H_2O + CO_2$
4 Colourless; milky (cloudy)

Page 140 Food additives

1 a Stop food from reacting with oxygen and turning bad
  b Tinned fruit; wine
  c 56 J
2 a To stop food spoiling
  b Packaging that changes the condition of the food to extend its shelf life
  c It prevents the need for additives such as antioxidants to be added to foods
  d A catalyst
  e An indicator shows how fresh a food is on the outside of a package; a central circle darkens as the product loses its freshness
3 a The tail is a 'fat-loving' part and the head is a 'water-loving' part; the fat-loving part of the molecule goes into the oil and attracts it towards this end; the water-loving part will not go in; the water-loving part stays out of the oil but is attracted to the water molecules; the oil is, therefore, 'hooked up' to the water
  b The mayonnaise does not separate as the egg yolk has a molecule that has two parts; one part is a water-loving part that attracts vinegar to it, called the hydrophilic head; the other part is a water-hating part that attracts oil to it, called the hydrophobic tail; the hydrophobic tail is attracted into the lump of oil but the head is not; the hydrophilic head is attracted to water and 'pulls' the oil on the tail into the water

Page 141 Smells

1 a i Acid + alcohol $\longrightarrow$ ester + water
  ii (Label to mixture in flask)
  iii (Label to upward condenser tube)
  iv At **X** the vapour is cooling down again and condensing back to a liquid
  v So that the mixture can be boiled/react for longer (without drying out)

condenser

X—

acid

alcohol

2

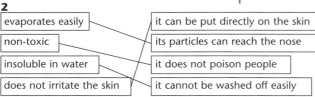

evaporates easily	it can be put directly on the skin
non-toxic	its particles can reach the nose
insoluble in water	it does not poison people
does not irritate the skin	it cannot be washed off easily

3 a Solution
  b Solvents
4 a Particles of a liquid are weakly attracted to each other; when some particles of a liquid increase their kinetic energy the force of attraction between the particles is overcome and the particles escape through the surface of the liquid into the surroundings; this is evaporation; if this happens easily the liquid is said to be volatile
  b This is because the force of attraction between two water molecules is stronger than that between a water molecule and a molecule of nail varnish; also the force of attraction between two varnish molecules is stronger than between varnish molecule and water molecule

Page 142 Making crude oil useful

1 a A molecule containing carbon and hydrogen only
  b i *(A: at the bottom, left-hand side, of the tower = 1 mark)*
  ii *(B: it 'exits' through the bottom of the tower = 1 mark)*
  iii *(C: at the top of the tower = 1 mark)*
  iv Fractions with lower boiling points such as petrol/LPG

A →

B

  c The forces between molecules are called intermolecular forces; these forces are broken during boiling/the molecules of a liquid separate from each other as molecules of gas; then either: the molecules in different fractions have different length chains; this means that the forces between the molecules are different; heavy molecules such as those that make up bitumen and heavy oil have very long chains; so they have strong forces of attraction between the molecules; this means that they are difficult to separate; a lot of energy is needed to pull each molecule away from another; they have high boiling points or: lighter molecules such as petrol have short chains; each molecule does not have very strong attractive forces and is easily separated; this means that less energy is needed to pull the molecules apart; they have very low boiling points *(Any 5 from either option)*
2 Oil slicks can harm animals, pollute beaches and destroy unique habitats for long periods of time; clean-up operations are extremely expensive and the detergents and barrages used cause problems
3 a C_7H_{16}
  b Alkenes have a double bond; alkanes have single bonds
  c Polymers
  d Cracking the fraction of heavy oil which is in excess supply to produce the smaller molecules needed for petrol which is in high demand but short supply

Page 143 Making polymers

1 a C
  b High pressure; catalyst
  c A double bond
  d

  '4 or 6 carbon atoms = 1 mark; alternate H and Cl atoms on bottom = 1 mark; brackets and bonds through either end = 1 mark; n at end = 1 mark)
  e

  H   CH$_3$
   \   /
    C==C
   /   \
  H   H

  (Two carbon atoms joined by a double bond = 1 mark; CH3 on top right hand side = 1 mark; only 4 other atoms/groups joined to two carbon atoms = 1 mark)
  f The reaction needs high pressure and catalyst; this causes the double bond in the monomer to break and each of the two carbon atoms forms a new bond; the reaction continues until it is stopped, making a long chain
2 a i It has an oxygen atom in its structure
  ii It contains a double bond
  iii A polymer made from the monomer butene
  b i This is because the bromine solution has reacted with the alkene and has formed a new compound
  ii Remains orange

Page 144 Designer polymers

1 a White dental filling is better than a mercury amalgam; waterproof plastics are better than fabric plasters

b i Hydrophobic means water-hating; the material repels water

   ii Water vapour from sweat can pass through the membrane but rainwater cannot so it keeps people dry when sweating; the membrane has pores which are 700 times larger than a water vapour molecule and therefore moisture from sweat passes through

2 a So that they do not have to be disposed of in landfill sites or burned but can decay by bacterial action

   b To make laundry bags for hospital so that they degrade when washed leaving the laundry in the machine
   (Or any other suitable use)

   c Landfill sites; waste valuable land; burning waste plastic: toxic gases; recycling: difficulty in sorting different polymers

3 a i *(See diagram)*   **ii** *(See diagram)*

intramolecular bonds

intermolecular forces

   b Some plastics have weak intermolecular forces of attraction between the polymer molecules so the polymer molecules can slide over one another/separate easily; some other plastics form intermolecular chemical bonds or cross-linking bridges between polymer molecules; these are strong so the polymer molecules cannot slide over one another; they are rigid/the chains cannot easily separate

Page 145 Using carbon fuels

1 a i Coal

   ii High energy value; good availability

   b Petrol and diesel are liquids so they can circulate easily in the engine; they are also stored easily in petrol stations along road networks; as these fuels are so easy to use and the population is increasing; more fossil fuels are being consumed, resulting in more carbon dioxide; this is a greenhouse gas; contributes to climate change which is a global problem
   (Any 4)

2 a Hydrocarbon fuel + oxygen $\longrightarrow$ carbon dioxide + water

   b i Carbon dioxide

   ii Water (steam) is tested by turning white copper sulphate to blue

   c Less soot is made; more heat is released; toxic carbon monoxide gas is not produced

   d People who live in the house are in danger of being made ill or even dying from carbon monoxide poisoning if the room is not well ventilated/heater faulty

   e $C_3H_8 + 5O_2 \longrightarrow 3CO_2 + 4H_2O$
   (Correct product formulae = 1 mark, correct balancing = 1 mark)

Page 146 Energy

1 a Exothermic; endothermic; exothermic; exothermic

   b Endothermic

   c Bonds are broken which is an endothermic reaction; new bonds are made which is an exothermic reaction; as less energy is needed to break bonds than make new bonds then a reaction is exothermic overall

2 Blue; complete; yellow; incomplete

3 a Measure the same mass of water in two beakers; put the burners under the beakers for the same time with the same rate of gas; measure the gas volume with a meter; measure the temperature of the water before and after the experiment/temperature increase; same mass water; same volume gas

   b i Same distance of the calorimeter from the flame; repeating the experiment 3 times and excluding draughts

   ii The energy transferred is calculated using the formula:
   energy transferred = mass of water x 4.2 x temperature change
   energy = 100 x 4.2 x 50
   = 420 x 50
   = 21000 J

   energy per gram = $\dfrac{\text{energy supplied}}{\text{mass of fuel burnt}}$

   = $\dfrac{21000}{4.00}$

   = 5250 J/g

P1 Energy for the home

Page 148 Heating houses

1 Energy flows from warm to cooler body; temperature of warmer body falls

2 Thermogram uses colours to represent different temperatures; car engine/tyres/exhaust will be hot; thermogram will show colours representing high temperature against cold field

3 a Specific heat capacity

   b Energy needed = mass x specific heat capacity x temperature change
   = 0.5 x 3900 x 70
   = 136 500 J

4 a Specific latent heat

   b Energy needed to break bonds; holding molecules together

Page 149 Keeping homes warm

1 a 120 ÷ 40 = 3 years

   b Shorter payback time

   c Only 32% of energy input is useful; as energy output

   d 0.32 x 6.5 = £2.08

   e Energy is lost up the chimney

   f i Solids

   ii Liquids and gases

   iii Radiation does not need a material

Page 150 How insulation works

1 a Particles in solid close together; gap between glass filled with gas or vacuum; particles in gas far apart/no particles in vacuum; more difficult to transfer energy than in solid
   (Any 3)

   b i Air in foam is good insulator; reduces energy transfer by conduction; air is trapped; unable to move; reduces energy transfer by convection *(Any 4)*

   ii Energy from room reflected back into room in winter; energy from Sun reflected back outside in summer

2 a Particles are in constant motion; particles vibrate and transfer kinetic energy – conduction; particles in solid close together so transfer energy easily; air is a gas so particles are far apart more difficult to transfer energy
   (Any 3)

   b Air expands when heated;

   density = $\dfrac{\text{mass}}{\text{volume}}$;

   increased volume means less dense

Page 151 Cooking with waves

1 a Microwave radiation is more penetrating than infrared; microwave ovens cook by conduction and convection
   (Any 1)

   b Microwaves need line of sight; no obstructions in space

   c Amplified; retransmitted back to Earth

2 a Gamma rays

   b Wavelength of radiation from iron longer than wavelength of radiation from element

3 Radio waves diffracted around hill; short wavelength/microwaves do not show much diffraction

Page 152 Infrared signals

1 **a** Digital

b

c Digital signals have only two states; interference not noticed; can multiplex digital signals

2 **a** **i** (x)   (y)   (z)

(diagram)

ii *(See diagram)*

b **i** No need for surgery; can see in real time

ii Light down one set of optical fibres; reflected from internal organs; up a second set of fibres; viewed by eyepiece/camera   *(Any 3)*

Page 153 Wireless signals

1 Less refraction at higher frequencies

2 Radio waves diffracted around mountain; shorter wavelengths do not show much diffraction; FM is shorter wavelength than LW

3 **a** Same as rotation time of Earth so appears to stay in same place

b 0.24 s

4 **a** The radio station is broadcasting on the same frequency; the radio waves travel further because of weather conditions

b Aimed at a very small object

Page 154 Light

1 **a** *(See diagram)*   **b** *(See diagram)*

(diagram showing waves with A and W labelled)

c Number of complete waves passing a point each second

2 **a** Need to represent letters as electric signal

b Advantage: signal received almost instantaneously; disadvantage: need line of sight; others can see signal *(Any 1)*

c White light: many colours; out of phase; laser light: one colour; in phase

Page 155 Stable Earth

1 **a**

description	P wave	S wave
pressure wave	✓	
transverse wave		✓
longitudinal wave	✓	
travels through solid	✓	✓
travels through liquid	✓	

b Waves refracted by core; cause shadow on opposite side of Earth

c S-waves do not pass through liquid; not detected on opposite side of Earth

2 **a** Reflects radiation back down to Earth

b Destroy ozone; reduce thickness of layer

c Ozone filters out ultraviolet radiation

d **i** Ultraviolet radiation; cells in skin produce melanin

ii Can stay in sun 30 times longer; without burning

B2 Understanding our environment

Page 157 Ecology in our school grounds

1 **a** **i** Too cold; no food

ii Difficult to reach so difficult to explore   *(Any 1)*

b **i** The range of living things in an ecosystem

ii Using pesticides kills the invertebrates; less food for the vertebrates; so their number also falls

(Any 2 = 1 mark each)

2 **a** $\frac{20 \times 10}{5} = 40$

b **i** Set more traps; over larger area

ii Larger sample size; be more representative of the whole wood not just a small area

Page 158 Grouping organisms

1 **a** **i** Birds   **ii** Beak; feathers

b No backbone

c

	food	shape	movement
animals	eat food	compact	move around

2 **a** Hybrid   **b** It is not a species

3 **a** Organisms that interbreed; produce fertile offspring

b **i** Bobcat; ocelot

ii First part of Latin name the same

Page 159 The food factory

1 **a** $6CO_2 + 6H_2O = C_6H_{12}O_6 + 6O_2$

b

product of photosynthesis	use in the plant
cellulose	cell wall
protein	growth/repair
oil	storage

2 **a** Carbon dioxide; temperature

b **i** As light intensity increases the rate increases; until point B when it levels off

3 **a** Respiration

b Need the energy it provides to live

Page 160 Compete or die

1 **a** The place an organism occupies within its ecosystem

b **i** Grey squirrels out-compete red

ii Increase in numbers

2 **a** **i** Decrease

ii More caterpillars; so more get eaten

b **i** Organism that feeds off another living organism

ii Mutualism

3 Population starts to rise because plenty of food; population starts to fall; because of increase in predator number; predator number falls so lemming population can rise again

(Any 3 = 1 mark each)

Page 161 Adapt to fit

1 **a**

adaptation	why it helps the camel survive
no fat on body, except in hump	whole body not insulated
hair-lined nostrils	stop sand getting into nose
higher body temperatures do not harm camel	higher body temperature reduces need to sweat

b **i** Layer of fat; for insulation; small ears; reduce heat loss; fur on soles; insulate feet

(Any 3 = 1 mark each)

ii Adapted to different habitats; cannot compete with brown bear in warmer habitat

2 **a** Reduces surface area; so reduced water loss

b Colourful petals; nectar; sticky pollen

(Any 1 = for 1 mark each)

c Pollen is smaller; feathery stigma; no nectar; no colour to flowers   *(Any 1 = for 1 mark each)*

Page 162 Survival of the fittest

1 a (B) C D A

(C before D = 1; D before A = 1; A at end = 1 mark)

b Some body parts decay and do not survive; some fossils have not been discovered yet; fossilisation is rare

c For: complex organisms appear then disappear/there are gaps in fossil record; against: there are examples of gradual change seen in the fossil record e.g. horse

2 a Organisms better adapted are more likely to survive

b More resistant bacteria survive; pass on resistance

Page 163 Population out of control?

1 a Climate change/greenhouse effect; climate change; melting ice caps; rise in sea level   *(Any 2 = 1 mark each)*

b i CFC   **ii** More ultra violet rays get through

2 a Exponential growth

b Developed countries use more fuel per person; developed countries have more cars; developed countries have more factories; developed countries have more fossil-fuelled power stations   *(Any 2 = 1 mark each)*

3 a Indicator species

b Mussels, damselfly lava and bloodworms survive in polluted water; if it was clean you would find alderfly and stonefly

Page 164 Sustainability

1 Loss of habitat; hunted; other animals out-competed it; disease; habitat polluted   *(Any 2 = 1 mark each)*

2 a Captive breeding programs/prevent hunting to extinction

b Tourism; improved transport; improved water supply/other resources   *(Any 2 = 1 mark each)*

3 a For: provide food/jobs; against: whales could become extinct

b Area too large to police

4 a Leaves enough fish behind to breed; maintain population

b Alternative to fossil fuels; can easily be replaced; because they grow fast   *(Any 2 = 1 mark each)*

C2 Rocks and metals

Page 166 Paints and pigments

1 a Made of tiny droplets of one liquid in water which is called an emulsion; when emulsion paint has been painted onto a surface as a thin layer; the water evaporates leaving the binding medium and pigment behind; as it dries it joins together to make a continuous film

b Oil paint and emulsion paints are colloids because they are a mixture of solid particles in a liquid; the particles/droplets are very small and stay scattered in the liquid; they do not settle out in the bottom; particles in paint are small enough to stay dispersed through the liquid while it is in use

c The solvent evaporates the oil slowly; reacts with oxygen in the air; the oil binding medium is oxidised by the air to form a tough, flexible film over the wood

2 a Used to paint cups; used to paint kettles; act as a warning   *(Any 2)*

b The thermochromic paints are mixed with different colours of normal acrylic paints in the same way that any coloured paints are mixed

c The mixture contains a blue thermochromic pigment and yellow acrylic paint; when the mixture gets hot the thermochromic pigment goes colourless; the green loses the blue colour so only the yellow of the acrylic paint is seen

d Phosphorescence

e Light

3 a In luminous clock dials

b Radioactive paints because the phosphorescent pigments are much safer

Page 167 Construction materials

1 a Limestone; marble; granite

b

building material	brick	cement	glass	iron	aluminium
raw material	clay	limestone and clay	sand	iron ore	aluminium

c Sedimentary

d Granite is an igneous rock made from molten rock which solidifies slowly forming interlocking crystals; marble is a metamorphic rock made by putting chalk/limestone under heat and pressure

e Limestone is made from the shells of dead sea-creatures that were compressed together

2 a Calcium carbonate $\longrightarrow$ calcium oxide + carbon dioxide   *(Each compound = 1 mark)*

b Limestone is heated; clay is added

c $CaCO_3 \longrightarrow CaO + CO$

d If a heavy load is put on a concrete beam it will bend very slightly; when a beam bends its underside starts to stretch; which puts it under tension and cracks start to form; steel is strong under tension; steel rods reinforce concrete to stop it stretching

Page 168 Does the Earth move?

1 a Less dense   **b** Continental plates; oceanic plates

c In plate tectonics, energy from the hot core is transferred to the surface by slow convection currents in the mantle; if two adjacent convection currents move clockwise/anticlockwise the plates will be pulled together; if anticlockwise/clockwise the plates will move apart; or they can scrape sideways past each other

plate   plate   plate   plate

convection current

d Two convection currents moving up and away from each other cause the plates to move apart; two convection currents moving down and towards each other cause the plates to collide; they can scrape sideways past each other.

e The coastlines of Africa and South America 'fit' suggesting that they have split and the continents have drifted apart; the sea floor shows ridges in the middle of the ocean suggesting plate movement

2 a Because it is less dense than the crust

b

small crystals	large crystals
cool rapidly	cool slowly
basalt	granite

c i Silica-rich magma

ii It shoots out as clouds of searingly hot ash and pumice; the falling ash often includes large lumps of rock called volcanic bombs; geologists investigate past eruptions by looking at the ash layers; in each eruption coarse ash falls first, followed by fine ash, producing graded bedding

Page 169 Metals and alloys

1 a It must first be analysed to find out how much of each element is present

b It has to be electrolysed again before it can be used

2 a Impure copper

b It is 'plated' with new copper

c It dissolves

d The cathode is plated in pure copper and gets thicker

e Blister copper/boulder copper

f They sink to the bottom of the cell

3 a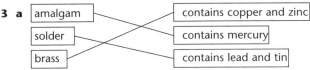

b Pure copper conducts electricity so well

c They can be bent more than steel so they are much harder to damage; they change shape at different temperatures, this is 'shape memory'

d In the frames of glasses to stop them breaking; in shower heads to reduce the water supply if the temperature gets so hot it scalds; a small piece of metal can be put into a person's blocked artery and then warmed slightly, as it warms up it changes shape into a much larger tube that holds the artery open and reduces the risk of a heart attack *(Any 2)*

e Nickel and titanium

Page 170 Cars for scrap

1 a Salt accelerates rusting which means that car bodies rust quicker

b It has a protective layer of aluminium oxide which does not flake off the surface

c It flakes off

d iron + water + oxygen ⟶ hydrated iron(III) oxide

2 a Iron; carbon

b Stronger, harder; does not rust as easily as pure iron *(Any 2)*

c Advantages: the mass of a car body made of aluminium will be less than the same car body made from steel; the car body made of aluminium will corrode less; disadvantage: the car body of the same car will be more expensive if made from aluminium

d

material and its use	reasons material is used
aluminium in car bodies and wheel hubs	less dense so greater fuel economy, corrodes less
copper in electrical wires	good electrical conductivity
plastic in dashboards, dials bumpers	easily cleaned, lightweight, not easily damaged
pvc in metal wire coverings	flexible, easy to colour
plastic/glass composite in windscreens	transparent, less easy to break
fibre in seats	soft, comfortable, flexible

3 More recycling of metals means that less metal ore needs to be mined; recycling of iron and aluminium saves money and energy compared to making iron from their ores; less crude oil is used to make plastics; less non-biodegradable waste from plastics is dumped; recycling batteries reduces the dumping of toxic materials into the environment *(Any 3)*

Page 171 Clean air

1 a i *(See diagram)*
ii *(See diagram)*
iii Combustion and respiration increase the level of carbon dioxide and decrease the level of oxygen; photosynthesis decreases the level of carbon dioxide and increases the level of oxygen

0.035% carbon dioxide
0.965/1% water vapour
21% oxygen

2 a

increased energy usage	more fossil fuels are being burnt in power stations
increased population	the world's energy requirements increase
deforestation	as more rainforests are cut down less photosynthesis takes place

b i The original atmosphere contained ammonia and later carbon dioxide

ii A chemical reaction between ammonia and rocks produced nitrogen and water; the percentage of nitrogen slowly increased; nitrogen is very unreactive so very little nitrogen was removed

iii Much later organisms that could photosynthesise evolved; these organisms converted carbon dioxide and water into oxygen; as the percentage of oxygen in the atmosphere increased, the percentage of carbon dioxide decreased, until today's levels were reached

3 a

pollutant	carbon monoxide	oxides of nitrogen	sulphur dioxide
origin of pollutant	incomplete combustion of petrol or diesel in car engine	*formed in the internal combustion engine*	*formed when sulphur impurities in fossil fuels burn*

b A rhodium catalyst

c i The two gases formed are natural components of air/no longer pollutants

ii Carbon monoxide + nitric oxide ⟶ nitrogen dioxide + carbon dioxide
(Reactants = 1 mark, product = 1 mark)

iii $2CO + 2NO \longrightarrow N_2 + 2CO_2$
(Reactants = 1 mark, products = 1 mark, balance = 1 mark)

Page 172 Faster or slower (1)

1 a i *(See graph)* **ii** The higher temperature
iii As the temperature increases the particles move faster; the reacting particles have more kinetic energy and so the number of collisions increases and the number of successful collisions increases

b The frequency of successful collisions

c It increases by the number of successful collisions between reactant particles that happen each second

2 a i The higher concentration of acid results in a quicker reaction rate

ii The same amount of hydrogen collected in both cases

b i $X = 10$ secs; $y = 25$ cm^3

Gradient $= \dfrac{y}{x} = 2.5$ cm^3/sec

ii Reading within the graph between closer points

Page 173 Faster or slower (2)

1 Carbon dioxide and water vapour; a large volume of gaseous products are released, moving outwards from the reaction at great speed, causing the explosive effect

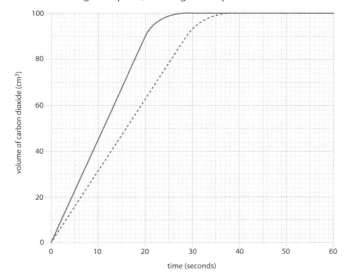

2 a A gas is given off: carbon dioxide
b i 27–29 seconds **ii** (See graph)
iii As the reactants are used up there are less frequent collisions
c There are more frequent collisions as there are more exposed particles
3 a It makes the reaction go faster
b It often provides a specific surface on which the appropriate chemicals can react

P2 Living for the future
Page 175 Collecting energy from the Sun

1 a Robust/not much maintenance; no fuel/no power cables; no pollution/no contribution to climate change; renewable energy source
b N-type has excess free electrons; p-type has absence of free electrons; impurity added
2 a Walls and floors radiate energy back into room
b i

X-rays	Ultraviolet	visible light	Infrared	radio
			S	**P**

ii
3 a Moving air
b Advantages: no atmospheric pollution; do not contribute to climate change; disadvantage: noisy; environmental impact; do not work if no wind; do not work if too windy *(Any 2)*

Page 176 Generating electricity

1 a Stronger magnet; more turns on coil; turn coil faster *(Any 2)*
b Magnetic field turns inside coil of wire
2 a Turbine turns generator
b Current in wire warms wire; loses energy to surroundings; higher voltage means less current; less energy loss *(Any 3)*
c Boilers; generators; cooling towers

Page 177 Fuels for power

1 a i Substance that burns
ii Releases energy as heat atoms of uranium split
b i Climate change
ii Renewable energy source

2 a Power = voltage x current
= 12 x 2
= 24 W
b Cost = power x time x cost per kWh
= 2.5 x 0.5 x 12
= 15p
c Not much is needed but it still has to be produced
3 a Nuclear; DNA; mutate; cancer
b Advantages: fossil fuels not being used up; no atmospheric pollution; no CO_2 leading to climate change; disadvantages: high maintenance costs; radioactive nuclear waste produced; risk of accident
(Any 4)

Page 178 Nuclear radiations

1 a Radon gas; rocks and soil; cosmic rays; food and drink *(Any 2)*
b False; true; true; false
c Negative ions: atom gains electrons; positive ions: atom loses electrons
2 a Tracer
b Penetrating
3 a Remains radioactive/harmful for that long
b Not suitable for making bombs; can be used to contaminate water supplies; can be used to contaminate areas of land

Page 179 Our magnetic field

1 a Magnetic field is similar to a bar magnet
b Solid inner core; moves within liquid outer core; comparison with dynamo effect
2 a The presence of the Moon
b Another planet in same orbit as old Earth; planets collided; iron core formed centre of new Earth; less dense rocks started to orbit; clumped together to form Moon *(Any 3)*
3 Spiral around magnetic field lines; cause auroras/lights in sky
4 Charged particles; radio signals

Page 180 Exploring our Solar System

1 a i (See diagram)
ii (See diagram)
b i Centripetal
ii

2 a Always gravitational forces acting however small
b Avoid being blinded by Sun
c Mars; nearest planet/atmosphere most similar to Earth
3 Distance travelled by light in one year

Page 181 Threats to Earth

1 a Mars; Jupiter
b Rocks left over when the Solar System formed
c Asteroids have low mass compared to Jupiter; gravitational force from Jupiter prevents asteroids joining together
2 a (C on elliptical orbit)
b (X on elliptical orbit at far left)
c Solar winds blow dust
3 a Increased accuracy of predicting if a collision may occur
b Send a rocket packed with explosives; explode rocket close to asteroid

Page 182 The Big Bang

1 a Those furthest away
  b Wavelength of light increases when light source moving away from observer; spectral lines on star's spectrum; move to red end of spectrum
  c Faster moving galaxies show more red shift
2 a Cloud pulled together by gravity; starts to get hot and glow; hot enough to produce nuclear fusion; joining of hydrogen atoms to form helium and release energy
  b i Red giant; core contracts; outer changes colour from yellow to red and expands; planetary nebula thrown out; core becomes white dwarf; cools to become black dwarf *(Any 4)*
    ii Red supergiant; core contracts; outer expands; core becomes neutron star; explosion/supernova; neutron star becomes black hole; supernova remnants become new stars *(Any 4)*

B3 Living and growing

Page 184 Molecules of life

1 Mitochondria
2 a James; pattern is identical
  b DNA isolated; DNA cut into fragment; fragments separated; by electric current/ electrophoresis; banding compared with a reference *(Any 3 = 1 mark each)*
  c i 5
    ii ATATAC ATTTTTGTT
3 a Biological catalyst; that speeds up reactions in body
  b As temperature increases the rate increases; until 40 °C when it falls
  c 40–42 °C/any temperature in this range
  d High temperature denatures enzyme; enzyme changes shape; so can no longer fit substrate

Page 185 Diffusion

1 a Movement of substances; from a region of high concentration to low concentration
  b Increasing surface area; decreasing diffusion distance; greater concentration difference
2 a Carbon dioxide; waste
  b Alveoli
3 Higher concentration of carbon dioxide in the air; diffuses through stomata; used in photosynthesis; which keeps concentration low in leaf *(Any 3 = 1 mark each)*
4 Releases transmitter substance; transmitter substance diffuses across gap; transmitter triggers impulse in next neurone

Page 186 Keep it moving

1 a Disc-shaped: large surface area takes up oxygen quicker; no nucleus: more space to carry more oxygen
  b Haemoglobin

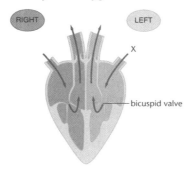

2 a i *(See diagram)*
    ii *(See diagram)*
  b Needs to pump blood further; at higher pressure
  c i Rejection; waiting for donor; need for anti-rejection drugs *(Any 2 = 1 mark each)*
    ii Advantage: no need to replace batteries; disadvantage: have to wait for suitable donor
3 a Arteries transport blood away from heart; capillaries allow the exchange of materials with tissue; veins transport blood back to heart
  b Blood going to body can be pumped at much higher pressure; provides greater flow to all organs

Page 187 Divide and rule

1 a Limits size organism can grow to; no cell differentiation; unable to form complex tissue e.g. nerves *(Any 2 = 1 mark each)*

  b i

cube	surface area in cm²	volume in cm³	ratio
A	24	8	24/8 = 3
B	54	27	54/27 = 2
C	96	64	96/64 = 1.5
D	150	125	150/125 = 1.2

    ii Larger cells have much smaller surface are compared to volume; difficult to absorb enough food/oxygen
2 a Mitosis
  b Cells are diploid; chromosomes separate to opposite poles
3 a Contains enzymes; to break down egg membrane
  b i Meiosis
    ii Homologous chromosomes separate; haploid cells form; variation occurs/new cells are different from each other; four cells produced *(Any 1 = 1 mark)*
  c Only contain one of each pair of chromosomes/ contain half the number of chromosomes found in body cells

Page 188 Growing up

1 a Similar: both have cytoplasm/cell membrane; different: plant cell has cell wall/vacuole/chloroplasts
  b i Stem cells
    ii Embryos die; unethical; religious beliefs *(Any 2 = 1 mark each)*

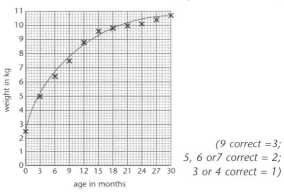

(9 correct =3; 5, 6 or 7 correct = 2; 3 or 4 correct = 1)

2 a i *(See diagram)*
    ii *(See diagram)*
    iii Weight increases; as age increases
    iv Infancy
  b Poor diet/problems with digestive system

Page 189 Controlling plant growth

1 a Stimulates root growth
  b i Speeds up growth too much; so they die
    ii Narrow leaves; they do not absorb as much weedkiller
2 a i Geotropism
    ii Because it grows in the same direction as gravity
  b i Auxin
    ii Produced in tip; more collects on the dark side; causes cell elongation on dark side
    iii No tip; so no auxin produced

Page 190 New genes for old

1 a Select characteristics; cross-breed; select the best offspring; breed these offspring over many generations *(Any 3 = 1 mark each)*
2 a TG just before AAA have swapped round to GTAA
  b Change in base sequence that codes for enzyme B; protein (enzyme) made is the wrong shape; red pigment cannot be turned into purple
3 a Find the gene for beta-carotene in carrots; remove gene; put gene into rice
  b Advantage: produces organisms with new characteristics; disadvantage: may have harmful effects

Page 191 More of the same

1 **a** Sperm collected from bull; cows artificially inseminated; embryo collected; embryo cloned; embryo implanted in surrogate cow *(Any 3 = 1 mark each)*

b i Pig organs used for transplants instead of humans; reduces need to wait for donors to die

ii Breeding pigs just to kill them; may not want an animal's organ; religious reasons *(Any 1 = 1 mark)*

c i Nucleus removed from body cell; no fertilisation takes place; clone of the adult *(Any 1 = 1 mark)*

ii B; because the nucleus came from cell from sheep B

2 **a** Advantage: characteristics all the same/mass produce plants quickly; disadvantage: susceptible to new disease/changes in the environment/lack variation

b Plant cell retains ability to differentiate; into all the different types of cells; easier to separate plant cells

C3 The periodic table

Page 193 What are atoms like?

1 **a** Protons; neutrons *(Both = 1 mark)*

b

	relative charge	relative mass
electron	–1	0.0005 (zero)
proton	+1	1
neutron	0	1

c The number of protons in an atom

d The total number of protons and neutrons in an atom

e Fluorine

f i 17

ii 18

2 **a** Isotopes are elements that have the same atomic number but different mass numbers

b

isotope	electrons	protons	neutrons
$^{12}_{6}C$	6	6	6
$^{14}_{6}C$	6	6	8

3 The first shell can only take 2 electrons; the second shell can only take up to eight; which is why a 3rd shell is needed

Aluminium

Page 194 Ionic bonding

1 **a** An atom which has extra electrons in its outer shell and needs to lose them to be stable

b

c i Positive

ii Magnesium/calcium

d i Gaining

ii Fluorine/chlorine/bromine/iodine

e Positive; negative; lattice

f $\left[Na \right]^{+} \quad \left[\times Cl : \right]^{-}$

g $\left[: Cl \times \right]^{-} \quad \left[Mg \right]^{2+} \quad \left[\times Cl : \right]$

h Sodium chloride solution; molten (melted) magnesium oxide; molten sodium chloride

i i High

ii Conduct electricity

iii Ions can move

Page 195 Covalent bonding

1 **a** Covalent bonding

b A molecule of water is made up of three atoms; two hydrogen and one oxygen; oxygen has six electrons in its outer shell; it needs two more electrons to be complete; hydrogen atoms each have one electron in their only shell; the oxygen outer shell is shared with each of the hydrogen electrons; so each of the hydrogen atoms has a share of two more electrons making the shell full

c Because they are covalently bonded

d i
H_2O

ii CO_2

2 They are simple molecules with weak intermolecular forces; they are easy to separate so the substances have low melting points

3 **a** The group number is the same as the number of electrons in the outer shell; it has 1 electron in the outer shell

b 7

c i 2

ii There are electrons occupying 2 shells

d 6

Page 196 The group 1 elements

1 **a i** Hydrogen gas

ii Their density is less than the density of water

b Sodium + water $\longrightarrow$ sodium hydroxide + hydrogen

c

reactivity increases down ↓		melting point in °C	boiling point in °C
	$_{3}Li$	179	1317
	$_{11}Na$	98	892
	$_{19}K$	64	774

d Group 1 metals all have 1 electron in their outer shell so they react in a similar way

e $2Na + 2H_2O \longrightarrow 2NaOH + H_2$

2 **a** They moistened a flame test wire with dilute hydrochloric acid; dipped the flame test wire into the sample of solid chemical; held the flame test wire in a blue Bunsen burner flame; put on safety goggles; recorded the colour of the flame in a table

b

3 **a** $Li - e^{-} \longrightarrow Li^{+}/Li \longrightarrow Li^{+} + e^{-}$

b Lithium loses its outer electron from its second shell; sodium loses its outer electron from its third shell; the third shell is further away from the attractive 'pulling force' of the nucleus so the electron from sodium is more easily lost than the electron from lithium; sodium is therefore more reactive than lithium

c Oxidation

Page 197 The group 7 elements

1 **a**

chlorine	green gas
iodine	grey solid

b They all have seven electrons in their outer shell

c i $Cl + e^{-} \longrightarrow Cl$

ii Reduction

iii Fluorine gains its last electron into its second shell; chlorine gains its last electron into its third shell; the third shell is further away from the attractive 'pulling force' of the nucleus so the electron to fluorine is more easily gained than the electron to chlorine; fluorine is therefore more reactive than chlorine

2 a Potassium + iodine $\longrightarrow$ potassium iodide

b $2K + I_2 \longrightarrow 2KI$

3 a i Chlorine displaces the bromide ions which become bromine solution which is red-brown/ a displacement reaction occurs

ii This is because chlorine is more reactive than bromine so bromine does not displace the chloride ions

b i Bromine + potassium iodide $\longrightarrow$ potassium bromide + iodine

ii $Br_2 + 2KI \longrightarrow 2KBr + I_2$

Page 198 Electrolysis

1 a The electrolyte is a dilute solution of sulphuric acid; two electrodes are connected to a dc source of electric current, between 6 V and 12 V, and placed into the electrolyte; the electrode connected to the negative terminal is the cathode; the electrode connected to the positive terminal is the anode; when the current is switched on bubbles of gas appear at both electrodes; water splits into two ions: H+ is the positive ion and OH⁻ is the negative ion; H+ is attracted to the negative cathode and discharged as hydrogen gas, H_2; OH⁻ is attracted to the positive anode and discharged as oxygen gas, O_2

b Because the formula of the compound breaking up is H_2O

2 a At the cathode: the water and sodium chloride split up into ions and the ions are free to move; the positive H+ and Na+ ions migrate towards the negative cathode; only the H+ ions are discharged; each H+ ion gains one extra electron from the cathode; a pair of atoms then bonds to become a molecule of hydrogen that forms part of the gas; $2H^+ + 2e^- \longrightarrow H_2$ *(Any 3)*

b The negative hydroxyl OH⁻ and Cl⁻ ions migrate to the positive anode; only the OH⁻ ions are discharged; four OH⁻ ions each gain one electron and combine to form an oxygen molecule and two water molecules; $4OH^- - 4e^- \longrightarrow 2H_2O + O_2$ *(Any 3)*

3 a The ore of aluminium oxide is bauxite; aluminium oxide is melted; aluminium is formed at the graphite cathode; oxygen is formed at the graphite anode; the anodes are gradually worn away by oxidation; this forms carbon dioxide; the process requires a high electrical energy input

b Aluminium oxide $\longrightarrow$ aluminium + oxygen

c At the cathode: electrons have been gained; this is an example of reduction; $Al^{3+} + 3e^- \longrightarrow 3Al$

at the anode: electrons have been lost; this is an example of oxidation; $2O^{2-} - 2e^- \longrightarrow 2O_2$

d Aluminium oxide requires large amounts of electricity to melt at very high temperatures, which is very expensive; cryolite lowers the melting point

Page 199 Transition elements

1 a i Copper compounds are blue

ii Iron(II) compounds are pale green

iii Iron(III) compounds are orange/brown

b i Iron is used in the Haber process to make ammonia

ii It is a transition metal, because it is in the transition metal block

c Copper carbonate $\longrightarrow$ copper oxide and carbon dioxide

2 a

ion	colour
Cu^{2+}	form a blue gelatinous solid
Fe^{2+}	form a grey/green gelatinous solid
Fe^{3+}	form an orange gelatinous solid

3 a $CuCO_3 \longrightarrow CuO + CO_2$

b $Fe^{3+} + 3OH^- \longrightarrow Fe(OH)_3$

Page 200 Metal structure and properties

1 a Lustrous; malleable

b Resistant to attack by oxygen/acids

c It has a low density

d *(It is close-packed positive metal ions; held together by a strong electrostatic force of attraction; between a 'sea' of delocalised electrons)*

electrons from outer shells of metal atoms are free to migrate

metal ions

e Because a lot of energy is needed to overcome the strong attraction between the delocalised electrons and the positive metal ions

2 a Materials that conduct electricity with little/no resistance

b Loss-free power transmission; super-fast electronic circuits; powerful electromagnets *(Any 2)*

c Superconductors only work at very low temperatures

d Because delocalised electrons within its structure can move easily

e i C

ii A

P3 Forces for transport

Page 202 Speed

1 a Average speed $= \dfrac{\text{distance}}{\text{time}}$

$= \dfrac{390}{3}$

$= 130$ km/h

b Car cannot maintain the same speed throughout

c Yes; if average speed is 130 km/h the car must have gone faster (and slower) than this at times

2 a i Distance = 5000 m

Time $= \dfrac{\text{distance}}{\text{average speed}}$

$= \dfrac{5000}{12.5}$

$= 400$ s

ii Average speed $= \dfrac{\text{distance}}{\text{time}}$

$= \dfrac{500}{35}$

$= 14.3$ m/s

iii Distance = 9 x 500

$= 4500$ m

Time = 400 − 35

$= 365$ s

Average speed $= \dfrac{4500}{365}$

$= 12.3$ m/s

b i D to E

ii Speed = gradient of OA

$= \dfrac{40}{40}$

$= 1$ m/s

iii Speed $= \dfrac{40}{80}$

$= 0.5$ m/s

Page 203 Changing speed

1 a

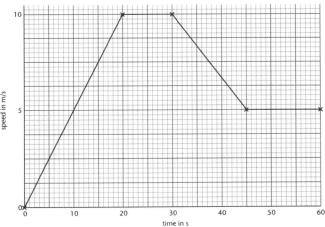

b Area under graph

c i

ii Yes; 50km/h = 50 000 m/h
$$= \frac{50\ 000}{3600}\ \text{m/s}$$
$$= 13.9\ \text{m/s}$$

iii Acceleration = gradient of graph from 0 to 15 s
$$= \frac{15}{15}$$
$$= 1\text{m/s}^2$$

iv Distance travelled = area under graph
$$= \tfrac{1}{2}(25 + 60) \times 15$$
$$= 637.5\ \text{m}$$

2 a
$$\text{Acceleration} = \frac{\text{change in speed}}{\text{time}}$$
$$= \frac{(40 - 10)}{6}$$
$$= 5\ \text{m/s}^2$$

b
$$\text{Acceleration} = \frac{\text{change in velocity}}{\text{time}}$$
Change in velocity = acceleration x time
$$= 6 \times 5$$
$$= 30\ \text{m/s}$$
Starts from rest so velocity after 5 s is 30 m/s

Page 204 Forces and motion

1 a
$$a = \frac{(v - u)}{t}$$
$$= \frac{(40 - 0)}{20}$$
$$= 2\ \text{m/s}^2$$

b F = ma
$$= 500 \times 2$$
$$= 1000\ \text{N}$$

c Resistive forces ignored

d
$$a = \frac{F}{m}$$
$$= \frac{1250}{500}$$
$$= 2.5\ \text{m/s}^2$$

e i (See diagram)
ii (See diagram)

iii Equal in size; opposite in direction; act on different bodies; mass of Earth is much greater than mass of Moon *(Any 3)*

2 a Tired; under influence of alcohol/other drugs; distracted/lacking concentration; icy/wet road; worn tyres; poor brakes *(Any 2)*

b i Increase
ii Thinking/reaction time unchanged; but she will travel a greater distance in that time at a higher speed

c Very little tread; less grip on road/car may skid

Page 205 Work and power

1 a WD = force x distance
$$= 80 \times 2$$
$$= 160\ \text{J}$$

b
$$\text{Distance} = \frac{\text{WD}}{\text{force}}$$
$$= \frac{160}{60}$$
$$= 2.67\ \text{m}$$

c i KE lost = braking force x braking distance
$$\text{braking distance} = \frac{\text{KE}}{\text{force}}$$
$$= \frac{300\ 000}{6\ 000}$$
$$= 50\ \text{m}$$

ii As heat; in tyres/road

2 a 60 x 10 = 600 N

b WD = force x distance
$$= 600 \times 3$$
$$= 1800\ \text{J}$$

c
$$\text{Chris' power} = \frac{\text{WD}}{\text{time taken}}$$
$$= \frac{1800}{8}$$
$$= 225\ \text{W}$$

d Priya's mass: WD = 225 x 10
$$= 2250\ \text{J}$$
2250 J = Priya's weight x 3
weight = 750 N
mass = 75 kg

3 Fuel pollutes the environment; car exhaust gases are harmful; carbon dioxide is a major source of greenhouse gases; carbon dioxide contributes to climate change *(Any 3)*

Page 206 Energy on the move

1 a Fewer road junctions, speed changes, gear changes *(Any 2)*

b $\frac{96}{24} = 4$

c Less

d Land Rover has a bigger engine capacity

2 a B; C

b i KE = $\frac{1}{2}$ mv^2
$$= \tfrac{1}{2}1200 \times (20)^2$$
$$= 240\ 000\ \text{J}$$

ii Loss in KE = braking force x braking distance force
$$= \frac{\text{KE}}{\text{distance}}$$
$$= \frac{240\ 000}{32}$$
$$= 7500\ \text{N}$$

iii No; KE proportional to v^2/quartered; so braking distance quartered, not halved

11

3 a Recharging requires electricity from power stations which cause pollution

b Energy from Sun so cause no pollution; do not have batteries that need recharging; do not use electricity from power stations *(Any 1)*

c Sun does not always shine; does not have a constant energy source *(Any 1)*

Page 207 Crumple zones

1 a

safety feature	how it works
seat belt	designed to stretch a little so that some of a person's KE is converted to elastic energy
crumple zones	absorb some of the car's KE by changing shape (crumpling) on impact
air bag	absorbs some of a person's KE by squashing up around them

b i Directly improve the safety of a car
ii Indirectly improve the safety of a car

c Traction control stops the wheels on a vehicle from spinning during rapid acceleration

d Active safety features have a more immediate effect; passive safety features contribute to good driving

e i Anti-lock braking system
ii Driver gets maximum braking force without skidding; can still steer car

2 a Force on people in car = mass x acceleration (F = ma); to reduce the force the deceleration (negative acceleration) must be reduced

b Crumple zones crumple up (or crush) slowing the car down more slowly

Page 208 Falling safely

1 a Golf ball heavier than ping pong ball; so air resistance has a bigger effect on ping pong ball

b 10 m/s^2

c i *(See diagram)*

air resistance

weight

ii Weight greater than air resistance

d The faster she falls the more air molecules she displaces each second; so the greater the air resistance force; net/resultant force is less so acceleration is less

e i Terminal speed
ii Balanced/equal in size but opposite in direction

f Larger; he reaches terminal speed when air resistance force = 1000 N; instead of 600 N in Sarah's case; so he must be moving faster before forces on him are balanced

g i Weight unchanged; upward force/air resistance suddenly increases by a large amount
ii Net upward force; so Sarah decelerates/slows down; air resistance force gradually reduces until it is again equal to Sally's weight; so she falls at a new, slower terminal speed

2 a Balanced
b Pull of engine/air resistance force
c Streamlined shape

Page 209 The energy of theme rides

1 a Energy converted into thermal energy and sound

b GPE = mgh
= 0.040 x 10 x 1.2
= 0.48 J

c When weight = air resistance force

2 a B
b C
c GPE to KE
d Energy transferred to other forms; e.g. sound/thermal energy (due to friction)
e Make B higher so that it gains greater GPE; its maximum GPE will be greater; GPE lost = KE gained + energy transferred to heat due to friction

3 a 65 x 10 = 650 N
b $\frac{400}{65}$ = 6.15 N/kg

B4 It's a green world

Page 211 Who planted that there?

1 a *(See diagram)*

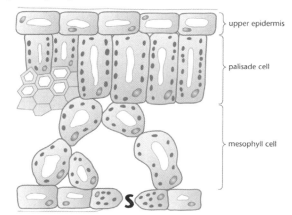

upper epidermis

palisade cell

mesophyll cell

b *(See diagram)*

2 Diffusion

3 a Thin so gases/light do not have far to travel; chlorophyll to absorb light; network veins for support/transport; stomata for gas exchange *(Any 3 = 1 mark each)*

b i Full of chloroplasts; to absorb a lot of light energy
ii Small and irregular; so they have a large surface area to volume ratio; for increased gas exchange *(Any 2 = 1 mark each)*

Page 212 Water, water everywhere

1 a Higher concentration water inside potato; water moved out

b Take on water; and burst (lysis)

c i Higher concentration of water outside cell; water moves into cell; increase in water pressure against cell wall; turgor pressure causes cell to be turgid *(Any two = 1 mark each)*

ii Cell drawn with cytoplasm shrinking away from wall

cell wall

cytoplasm

iii Plasmolysed

2 a Photosynthesis; cooling; support *(Any 2 = 1 mark each)*
b Photosynthesis stops in dark; less sugar produced; water concentration in guard cell increases; water moves out by osmosis *(Any 3 = 1 mark each)*

Page 213 Transport in plants

1 a A: phloem; B: xylem
   b Transport water/minerals; from root to leaves; support
     plant   *(Any 2 = 1 mark each)*
   c Contains lignin; for strength/cells die to make tubes; to
     carry water through
2 a Any value below 7.4 g
   b i Lowers rate
     ii Increases humidity; which reduces evaporation

Page 214 Plants need minerals too

1 a

mineral	why the mineral is needed	how the mineral is used
phosphate	respiration or growth	used to make cell membranes or DNA
potassium	respiration or photosynthesis	make enzymes
magnesium	photosynthesis	make chlorophyll

   b i Phosphate: poor root growth; discoloured leaves
     ii Potassium: poor flower or fruit growth; discoloured
       leaves
     iii Magnesium: yellow leaves
2 a i 0.0825
     ii 0.01/0.0125
   b Because there is a greater concentration in the cells;
     compared to the pond water
   c Active transport
   d Provide energy; needed for active transport

Page 215 Energy flow

1 a Numbers of organisms at each stage of food chain
   b

mass of 1 fox
mass of 100s of shrews
mass of 1000s of caterpillars
mass of 1 oak tree

   c Heat from respiration/egestion of waste/movement
2 a Use seeds for next year; use as biofuel
   b i 3090 – 1909 – 1022 = 159 kJ
     ii $\frac{159}{3090}$ × 100 = 5.1%
   c More energy efficient
3 a Yeast is used to ferment the sugar; to produce alcohol;
     alcohol is mixed with petrol   *(Any 2 = 1 mark each)*
   b Advantage: renewable energy source/cause less
     pollution/help countries become self reliant;
     disadvantage: can produce sulphur dioxide

Page 216 Farming

1 a DDT gets into plankton; passed along food chain; does
     not breakdown; builds up to toxic levels in seals
                                          (Any 3 = 1 mark each)
   b Less heat loss; less energy loss from movement
2 a By hydroponics; roots grow in water containing correct
     nutrients
   b Advantage: control mineral supply/conditions/disease so
     lower cost; disadvantage: manufactured fertiliser cost/tall
     plants need support
3 a Use of animals manure/compost; nitrogen fixing crops;
     seeding; growing seeds at different times
                                          (Any 2 = 1 mark each)
   b Cannot grow enough crops
   c i Biological control
     ii No predators so population keeps increasing

Page 217 Decay

1 a i 6.2 – 5.8 = 0.4
     ii Microorganisms had been killed
   b Breaks up remains; increasing surface area
   c i Bag C 40 °C
     ii Optimum temperature; for enzymes involved
   d Release enzymes onto food; enzymes digest food; fungi
     absorb digested food   *(Any 2 = 1 mark each)*
2 Kills bacteria/fungi by removing water

Page 218 Recycling

1 a i: burning; ii: respiration; iii: photosynthesis
   b Photosynthesis
   c Carry out respiration; when they break down the dead
     plant and animal material
   d Turn into limestone; limestone weathered by acid rain;
     chemical reaction produces
     carbon dioxide
2 a To make proteins for growth
   b Break down dead plants and animals; turning nitrogen
     compounds into nitrates
   c Unreactive
   d Nitrogen fixing bacteria; lightning
   e Denitrifying

C4 Chemical economics

Page 220 Acids and bases

1 a i An alkali is a base which dissolves in water
     ii Acid; water
   b Copper carbonate + sulphuric acid ⟶ copper
     sulphate + water + carbon dioxide
   c The salt formed is sodium nitrate
   d i NaCl
     ii $2HCl + CaCO_3 \longrightarrow CaCl_2 + H_2O + CO_2$
2 a Hydrogen ions, H^+
   b Hydroxide ions OH^-
   c Water H_2O
3 a The pH at the start is low as alkali is added the pH
     increases
   b The colour starts as purple. The pH falls as the acid
     neutralises the alkali; colour changes to blue; when
     neutral, the pH = 7; Colour is green

Page 221 Reacting masses

1 a 23 + 16 + 1 = 40
   b 40 + 12 + (16 × 3) = 100
   c 40 + 2 (16 + 1) = 40 + (2 × 17) = 74
   d When chemicals react, the atoms of the reactants 'swap'
     places to make new compounds – the products; these
     products are made from just the same atoms as before;
     there is the same number of atoms at the end as there
     were at the start, so the overall mass stays the same
2 a i 28
     ii 42
     iii Actual yield × 100 = percentage yield
     iv $\frac{28 \times 100}{42}$ = 66%
   b i $ZnCO_3 \longrightarrow ZnO + CO_2$
     ii Relative formula mass of $ZnCO_3$ is 65 + 12 + 16 + 16
       + 16 = 125;
       relative formula mass of CO_2 is 12 + 16 + 16 = 44;
       so if 125 g ZnCO3 gives 44 g CO_2;
       then 12.5 g ZnCO3 gives $\frac{12.5}{125}$ × 44
                              = 4.4 g CO2

Page 222 Fertilisers and crop yield

1 a To increase their crop yields
   b They are dissolved in water so they can be absorbed by
     plants through their roots
   c It is needed to make plant protein for growth.

2 a $(NH_4)_2SO_4$
$Mr = 2(14 + 4) + 32 + (16 \times 4) = 132$

b The relative formula mass of KNO3 is $39 + 14 + (16 \times 3)$ = 101;
mass of nitrogen = 14;
the percentage that is nitrogen is $\frac{14}{101} \times 100 = 13.8\%$

3 a Water
b i Phosphoric acid
ii Ammonium hydroxide
iii Phosphoric acid + ammonium hydroxide $\longrightarrow$ ammonium phosphate + water

c Phosphoric acid is reacted with ammonium hydroxide; the amounts used in the reaction must be exactly right, so a titration is carried out; titrate the alkali with the acid, using an indicator; repeat the titration until three consistent results are obtained; this is a neutral solution of potassium nitrate fertiliser, but it is contaminated with indicator; use the titration result to add the correct amounts of acid and alkali together without the indicator; the fertiliser made is dissolved in water, so most of the water is evaporated off using a hot water bath; leave the remaining solution to crystallise, then filter off the crystals

4 Eutrophication is caused by fertiliser run off increasing the amount of algae; this algal bloom prevents sunlight from reaching the plants below the water; anaerobic bacteria use up the oxygen in the water so fish die

Page 223 The Haber process

1 a Nitrogen is obtained from the air; hydrogen comes from natural gas; the gases are passed over an iron catalyst; under high pressure; an optimum temperature of 450 °C is chosen; there is a recycling system for unreacted nitrogen and hydrogen *(Any 3)*
b i A higher pressure increases the percentage yield but high pressures costs more
ii The high temperature decreases the percentage yield; however, higher temperatures make the reaction go faster
iii 450 °C is an optimum temperature; the yield is not as good, but that yield is made faster, so a satisfactory amount is produced in the right time

2 a 400 atmospheres
b Increases
c Decreases
d High temperature means higher rates but lower yields so it runs at the optimum temperature of 450 °C; this temperature means higher energy costs and also lower yields, but the increase in rate compensates; the plant produces more ammonia in a day at this temperature than it would at lower temperatures; total energy costs are not only due to heating costs; the plant needs compressors and pumps to achieve a high pressure; high pressure costs more so a lower, optimum pressure is used; although the reaction has a low percentage yield, the unreacted chemicals are recycled, and can go back into the reaction vessel, saving costs *(Any 3)*

Page 224 Detergents

1 a Organic acid + alkali $\longrightarrow$ detergent (salt) + water
b It is suitable for cleaning uses because; it dissolves grease stains; it dissolves in water at the same time *(Any 2)*
c i It is better to wash clothes at 40 °C instead of at high temperatures because washing machines have to heat up a lot of water; this needs energy; the lower the temperature of the water the less energy is used and less greenhouse gases are put into the atmosphere
ii As many dyes are easily damaged by high temperatures; it also means that many more fabrics can be machine washed as their structure would be damaged at higher temperatures

d The hydrophobic part of the molecule goes into the grease; 'hydrophobic' means water-fearing; the molecule forms bonds with the oil or grease; the hydrophilic part of the molecule forms bonds with the water; 'hydrophilic' means water-loving; the molecule forms bonds with the water and 'pulls' the grease off the fabric/dish into the water; the grease is pulled off the surface by the hydrophilic part forming bonds with water and lifting it away

2 a Intermolecular forces
b

covalent bond
hydrogen bond

Page 225 Batch or continuous?

1 a New batches are made when the stored medicine runs low; if a lot of one medicine is needed, several batches can be made at the same time; once they have made a batch of one drug it is easy to switch to making a different drug *(Any 1)*
b The process keeps going all the time so can be automated more easily
c Because it works at full capacity all the time it costs an enormous amount to build, but once running, it makes a large amount of product; employs very few people; making the cost per tonne very small
d They are flexible; it is easy to change from making one compound to another

2

strict safety laws
research and development
raw materials
labour intensive

The medicines are made by a batch process so less automation can be used.

They may be rare and costly

They take years to develop

People need to be feel a benefit without too many side effects

3 There has to be an anticipated demand for the drug and a potential market for it; there will be huge research and development costs as the time taken is often up to ten years; it is a very expensive process as labour costs are high; promising compounds often have dangerous side-effects so lots of similar compounds have to be made to find the best one with the fewest side-effects; tests can be made on as many as 10,000 compounds to find one effective drug; a new drug must be trial tested for safety, eventually using human trials and then submitted for approval; only then can it start earning money; the company patents the drug, so for the next 20 years or so they can sell it for a high price; the drugs company has to recoup all its development costs in that time as they have to cover the initial investment over the pay-back period; after that other companies can make their own version of the drug; these are called generic drugs *(Any 6)*

Page 226 Nanochemistry

1 a

diamond	[buckminster fullerene structure image]
graphite	[graphite layered structure image]
buckminster fullerene	[diamond tetrahedral structure image]

(matching lines cross-connecting diamond, graphite, buckminster fullerene to structures)

b

	diamond	graphite
use	cutting tools/ jewellery	electrodes/pencil lead/lubricant
reason	very hard/lustrous and colourless	conduct electricity/ high melting point

c

diamond	graphite
Does not conduct electricity because it has no free electrons	Slippery because layers of carbon atoms are weakly held together and can slide easily over each other

2 a Allotropes
b In diamond, each atom is held by covalent bonds to four other atoms, tetrahedrally, which are bonded further in different directions; this is called a giant structure; so many strong covalent bonds make the diamond hard; the bonding results in no free electrons, so it does not conduct electricity

3 a Semi-conductors in electrical circuits; industrial catalysts; reinforcement of graphite in tennis rackets   *(Any 2)*
b There is a very large surface area available

Page 227 How pure is our water?

1 a Sedimentation; filtration; chlorination
b Sedimentation: chemicals are added to make solid particles/bacteria settle out/larger bits drop to the bottom; filtration: a layer of sand on gravel filters out the remaining fine particles; some types of sand filter also remove microbes; chlorination: chlorine is added to kill microbes
c i Distillation is used to remove the dissolved substances
ii Distillation uses huge amounts of energy and is very expensive; it is only used when there is no fresh water

2 a Clean water saves more lives than medicines; that is why, after disasters and in developing countries, relief organisations concentrate on providing clean water supplies
b All water is recycled around the planet; if there is not enough rain in the winter, reservoirs do not fill up properly for the rest of the year

3 a Lead nitrate + potassium chloride ⟶ lead chloride + potassium nitrate
b Silver nitrate + potassium bromide ⟶ silver bromide + potassium nitrate
c $AgNO_{3(aq)} + KBr_{(aq)} \longrightarrow AgBr(s) + NaNO_{3(aq)}$

P4 Radiation for life
Page 229 Sparks!

1 a i So that charge cannot pass through her
ii So that she does not get an electric shock
iii Sally becomes charged
iv All her hairs gain the same charge; like charges repel; so the hairs move away from each other
b Each ball has the same charge; like charges repel
c Electrons in uncharged ball are repelled from the negatively charged ball; this leaves the side of the uncharged ball nearest the charged ball positively charged; opposite charges attract so the balls move towards each other

2 a i The car becomes charged due to friction with the air on the journey; you are not charged so charge flows through you when you touch the car door
ii Lightning may strike the tree as it is the tallest object around so you can become charged if you are under the tree
iii Cling film becomes charged due to friction as it is unrolled (as electrons are transferred from one part of the film to another, areas acquire opposite charges, so attract)
b Due to friction of the moving parts
c So that charge cannot flow through them to earth giving them an electric shock

Page 230 Uses of electrostatics

1 a Makes the heart contract
b Place paddles firmly on chest; no clothes/hairs
c It only passes for a very short time
d 400 J

2 a Paint droplets all have the same charge; like charges repel
b To attract the paint droplets to the frames; as opposite charges attract
c i Positive
ii Further paint droplets are repelled by the positive charge on the frame

3 a Positive
b Negative

Page 231 Safe electricals

1 a i Varies the brightness
ii Ammeter in series; voltmeter in parallel with lamp
iii $R = \dfrac{V}{I}$
$= \dfrac{6}{0.25}$
$= 24$ ohms
b i $R = \dfrac{12}{0.4}$
$= 30$ ohms
ii Electrons bump into atoms of the resistor giving them energy; atoms vibrate more; resistor gets hotter; its resistance increases

2 Battery is dc, mains is ac; battery is lower voltage than mains

earth

E

L

F

neutral   N   live

3 a i *(Live: on right (to fuse); neutral: on left; earth: on top)*
  ii Earth
  iii Earth wire is connected to metal case of an appliance; if a fault occurs and the case becomes live any charge on it flows to the ground via the earth wire; it provides a low resistance path to the ground
b i 13 A
  ii To break the circuit before current reaches the appliance
  iii Earth wire is connected to metal case of an appliance; if the case becomes live a large current flows in the earth and live wires; and fuse 'blows'

Page 232 Ultrasound
1 a i Vibrations set up pressure wave in air; compressions (higher pressure) and rarefactions (lower pressure); make eardrum vibrate
  ii Frequency increases
b Sound of a higher frequency than humans can hear
c Longitudinal: vibrations in same direction as wave travels; transverse: vibrations perpendicular to direction in which wave travels
2 a Pulse; tissues; reflected; echoes; image; gel; probe; skin; ultrasound/pulse; reflected; skin
b Densities of adjoining tissues; speed of sound in adjoining tissues
c i Its density needs to be similar to that of air so that nearly all the ultrasound is transmitted into the body at its surface (without the gel nearly all the ultrasound would be reflected at the surface of the body)
  ii 1600 m/s
3 a Very rapid ultrasound vibrations break the stones down into small pieces that are excreted from the body in the normal way
b It needs to be powerful enough/carry enough energy; to break up the stones

Page 233 Treatment
1 a Both very penetrating/can pass into the body to treat internal organs
b Alpha particles cannot penetrate skin; beta particles would be stopped by a small thickness of tissue and by bone
c Nucleus decays; emits alpha or beta particle; any surplus energy is emitted as gamma radiation
2 a A tiny amount of a radioscope introduced into the body
b To investigate a problem without surgery
c Gamma
d Thyroid
e X-rays are produced in an X-ray tube; gamma rays can be emitted inside the body and their progress monitored
3 a i Gamma
  ii Alpha particles cannot penetrate the skin; beta particles would be stopped by a small thickness of tissue and by bone
b So that the tumour receives the required dose of radiation to destroy it but healthy tissue only receives one third of the amount
c Rotate the source around the patient's head in a circle centred on the tumour

Page 234 What is radioactivity?
1 a

type of radiation	charge (+, – or 0)	what it is	particle or wave
alpha	+	helium nucleus	particle
beta	–	electron	particle
gamma	0	electromagnetic radiation	wave

b Gamma; alpha; alpha; gamma; beta
2 a The average time taken for half the nuclei in a sample of radioisotope to decay
b 5.5–5.9 years
3 a 86
b 134
c Alpha particle
d i $^{220}_{86}Rn \longrightarrow {}^{216}_{84}Po + {}^{4}_{2}He$
  ii 220 = 216 + 4
    86 = 84 + 2
e A neutron changes into a proton and an electron; the electron is emitted from the nucleus at high speed – beta particle
f $^{131}_{53}I \longrightarrow {}^{131}_{54}Xe + {}^{0}_{-1}b$

Page 235 Uses of radioisotopes
1 a Radioactive substances present in rocks (especially granite) and soil, cosmic rays from space
b Waste products from nuclear power stations; waste products from hospitals; man-made radioisotopes; nuclear weapons testing   *(Any 2)*
2 a Gamma; alpha and beta radiation would not penetrate pipe and soil to reach surface
b Geiger-Muller tube
c Count rate recorded will rise above the leak
3 a It is highly ionising/short range in air
b Alpha particles ionise atoms in the air; + and - ions move towards - and + plates respectively; this creates a tiny current which is detected; if present smoke particles attach themselves to the ions neutralising them; current falls setting off alarm
4 a Very little change in count rate over 200 year period
b Seeds; animal bone
c Rocks are not made from living matter
d 4000 years

Page 236 Fission
1 a Source of energy; water; steam; steam; turbine; generator
b The splitting of a large nucleus such as uranium with the release of energy
c Uranium containing a much higher proportion of the uranium-235 isotope than occurs naturally
d i Nucleus becomes unstable/splits; forming two daughter nuclei; extra neutrons; and energy
  ii Extra neutrons can go on to split more uranium nuclei; emitting even more neutrons
2 a Embedded in glass discs and buried in the sea/incinerated under strict controls (very low waste only)
b Reprocessed
3 a i Slows down the neutrons
  ii To increase the likelihood of a neutron being captured by a uranium nucleus
  iii Boron absorbs neutrons; boron rods raised and lowered in the reactor to alter the number of neutrons available for fission
b Advantage: less pollution/does not produce acid rain/only a small amount of fuel needed/larger reserves of uranium than of fossil fuels; disadvantage: dangerous waste products/waste products radioactive for a long time/decommissioning problems/risk of nuclear accident
  (Any 1 of each)